FREDERICK DU CANE GODMAN

NATURAL HISTORY

OF

THE AZORES

OR

WESTERN ISLANDS

Elibron Classics
www.elibron.com

Elibron Classics series.

ISBN 1-4021-6529-3 (paperback)
ISBN 1-4021-2533-X (hardcover)

This Elibron Classics Replica Edition is an unabridged facsimile of the edition published in 1870 by John van Voorst, London.

NATURAL HISTORY

OF

THE AZORES,

OR

WESTERN ISLANDS.

BY

FREDERICK DU CANE GODMAN, F.L.S., F.Z.S.,
ETC.

LONDON:
JOHN VAN VOORST, PATERNOSTER ROW.
MDCCCLXX.

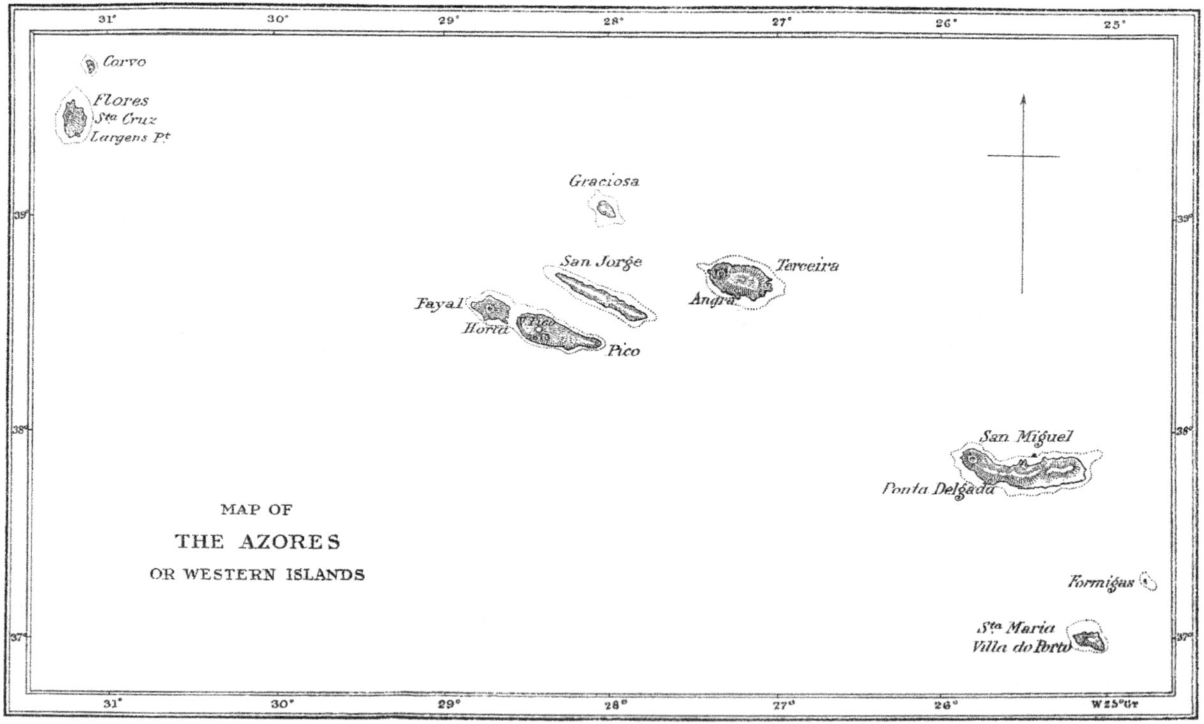
MAP OF
THE AZORES
OR WESTERN ISLANDS
Corvo
Flores
Sta Cruz
Largens Pt
Graciosa
San Jorge
Tereeira
Angra
Fayal
Horta
Pico
San Miguel
Ponta Delgada
Formigas
Sta Maria
Villa do Porto
31°
30°
29°
28°
27°
26°
25°
W 25° Gr
39°
38°
37°

NATURAL HISTORY

OF

THE AZORES,

OR

WESTERN ISLANDS.

BY

FREDERICK DU CANE GODMAN, F.L.S., F.Z.S.,

ETC.

LONDON:

JOHN VAN VOORST, PATERNOSTER ROW.

MDCCCLXX.

ALERE FLAMMAM.

PRINTED BY TAYLOR AND FRANCIS,

RED LION COURT, FLEET STREET.

PREFACE.

In the following pages I have endeavoured to give the result of a personal investigation of the Fauna and Flora of the Azores, made during a four months' visit to the archipelago. The materials accumulated during my stay consisted of numerous specimens, illustrating nearly every branch of Natural History. Since my return, these have been placed in the hands of several naturalists for determination. In their various reports, which constitute a large portion of this volume, all that was previously known relating to the Natural History of these islands has been incorporated, thus bringing our knowledge of the subject, in a complete form, down to the present time.

The animal and vegetable life existing in a group of islands like the Azores, separated by a wide expanse of deep sea from any other land, possesses a special interest. It is by the investigation of the indigenous products of such isolated spots that the greatest amount of light is to be thrown on the important subjects of the origin and distribution of species. It was with the view of increasing our knowledge on these points, that, five years ago, I visited the Azorean Archipelago.

No complete biological account of any island or group of islands exists at the present time, so far as I am aware, in a connected form. This volume is an attempt to supply the deficiency with regard to the Azores. That it falls far short of completeness I am fully conscious. It would require years of accurate observation, carried on through all seasons in the islands themselves, to render a work complete; it would require, too, a more extended knowledge than I possess.

To the naturalists who have cooperated with me, my thanks are especially due; for without their assistance it would have been impossible for me to undertake my task. I must mention Mr. Crotch, who, having investigated with Mr. Wollaston the Madeira and the Canary Islands, was thus thoroughly competent to undertake the determination of a collection of Coleoptera from the Azores. Mr. Tristram, whose wide knowledge of Terrestrial Mollusks is well known, has contributed some valuable notes. It has been my especial good fortune to enlist the interest of Mr. Watson in my undertaking. The Botany of the islands, his study of which began during a personal visit, has continued for many years to engage his attention; and thus he was better qualified than any other botanist to deal with this portion of the subject. Mr. Mitten's article will doubtless be acceptable as a valuable contribution to our knowledge of the Mosses and Hepaticæ of the Atlantic Islands in general. Dr. Günther, Mr. Frederick Smith, and Mr. Stainton have also materially assisted me in the departments of science in which they are especially proficient. It is with pleasure, too, that I here acknowledge the general help I have throughout received from Mr. Osbert Salvin, both in the arrangement and completion of my undertaking.

In conclusion, I have only to add that, in the "General Remarks," I can lay claim to no originality in the views I have there put forward. I have merely endeavoured to reason upon the facts before me as independently as possible; nearly the whole, indeed, was written before I had read the chapter on "Insular Floras and Faunas" in the tenth edition of Sir Charles Lyell's 'Principles of Geology.' The same facts have suggested a similar line of argument, and indicated the same general conclusions. This subject of Insular Floras and Faunas will be found more fully treated in the chapter to which I have already referred and in Dr. Hooker's most instructive lecture on Insular Floras, delivered before the British Association for the Advancement of Science at Nottingham, in August 1866.

To each article the Author's name is prefixed; where, however, no name appears, I must be considered responsible for what follows.

F. DU C. GODMAN.

Sept. 1870.

CONTENTS.

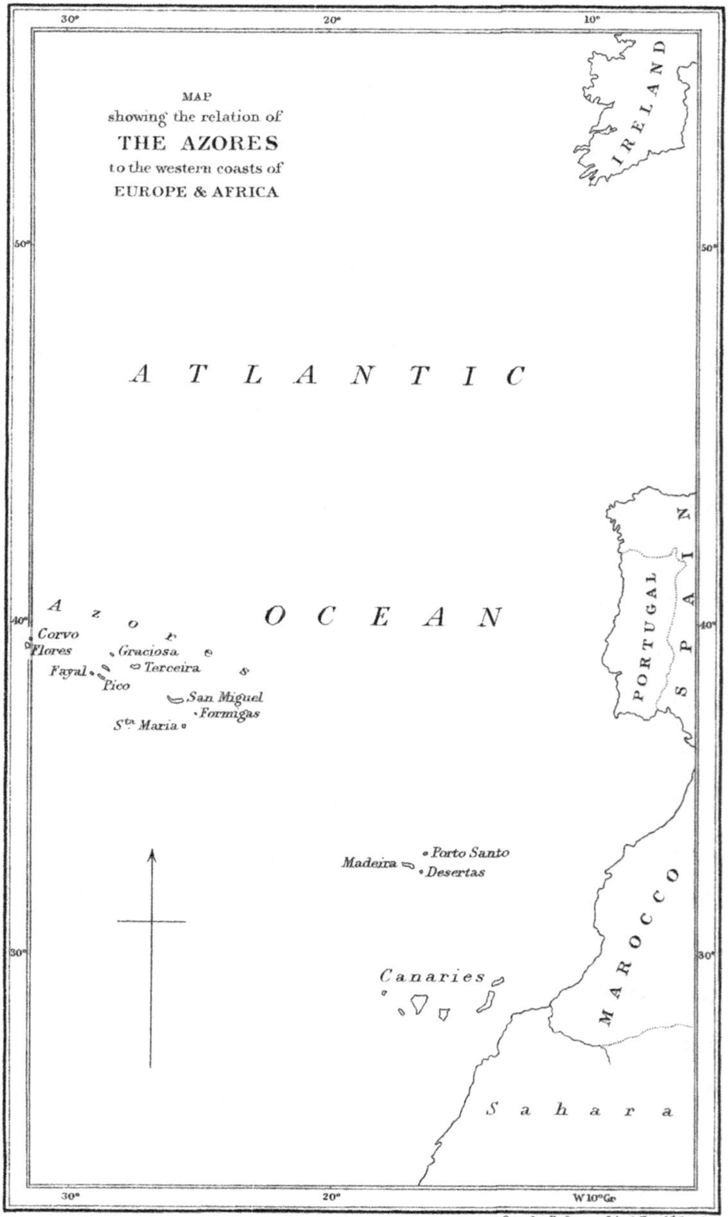

Stanford's Geographical Estabt. London

NATURAL HISTORY

OF

THE AZORES.

NARRATIVE.

Early one morning in August 1861 I passed through the Azores, on board the Royal Mail-steamer 'Seine,' on my outward passage from Southampton to the West Indies. The sea was perfectly calm; but the atmosphere was dense and oppressive, and heavy clouds rested on the top of Pico. I almost wished I could then have stopped and gone ashore to visit these isolated islands; but onward we went, and by breakfast-time were again out of sight of land. Sombrero, the most eastern of the West-Indian islands next came in view. Several times during my voyage I thought of the Azores, and made up my mind that, should the opportunity offer, I would visit them. I was absent from England some months; and on my return my mind was filled with the wonders of a tropical climate; I had almost forgotten the Azores; and it was not till the spring of 1865 that I was able to carry out this long-wished-for project. I was anxious to enjoy their scenery, as well as to investigate their fauna and flora.

The careful researches of Mr. Wollaston and others have brought to light numerous and very interesting forms amongst the Coleoptera of Madeira, the Canaries, and the Cape-Verde Islands. The Azores, however, remained very imperfectly explored*; and it was with the view of giving a more satisfactory *résumé* of the natural productions of this latter group, and to trace the relationship they bear to the other archipelagos, that I undertook the expedition.

On the 9th of March 1865, I and my brother, Captain Godman, started from Southampton, on board the Brazilian Mail Steam-vessel 'Oneida,' and on the 13th we landed at Lisbon. Here we found that the 'Leal,' a small screw-steamer, was to sail for the Azores on the 15th. This vessel runs, with more or less regularity, once a month, and, calling at five of the principal islands, returns direct to Lisbon. These two days we employed in visiting Cintra and its neighbourhood; and at 4 o'clock on the appointed day we found ourselves on board, steaming down the Tagus. We had few fellow-passengers, and but a light cargo; heavy weather, however, delayed our arrival; and it was not till the morning of the 21st that we anchored in the open roadstead of Ponta Delgada, the capital of St. Michael's. The gales which had followed us on our voyage were now succeeded by a perfect calm, leaving heavy clouds resting on the tops of the higher mountains, which, together with the dark foliage of the orange-groves and native evergreens, gave the island a peculiarly gloomy appearance.

It was nearly the end of the orange-season, but there were still about a dozen English schooners anchored off the town waiting for cargoes; and at a short distance out at sea were two more trying to come in, on board one of which was my collector, Mr. Brewer, whose services I had engaged to assist me more particularly in the Coleoptera.

* In another place will be found what had been already done by Messrs. Morelet, Drouet, Watson, and others, in their several departments of natural history.

He had sailed direct from England, and had also experienced heavy weather, though on the whole he had made a fairly prosperous voyage of a fortnight from London.

The Azores lie between long. 25° and 30° 15′ W., and the most eastern islands are 16° west of Lisbon. They are nine in number, and may be divided into three groups—St. Michael's and St. Mary's forming the eastern, Terceira, Graciosa, St. George's, Pico, and Fayal the central, and Flores and Corvo the extreme western. They have an agregate area of 700 square miles, and are of volcanic origin.

St. Michael's, the largest, and to us the best known, from its trade in oranges, has a population of about 80,000 inhabitants. It stretches nearly east and west, being much longer than it is broad. At the eastern end the mountains rise to a height of upwards of 3500 feet, and are chiefly covered with tree-heath (*Erica azorica*), Juniper (*Juniperus oxycedrus*), Faya (*Myrica faya*), and other evergreen shrubs. The peak of Agua de Poa in the centre reaches a height of 3070 feet; between this and the west end the land is lower, but still studded with numerous small volcanic cones, nearly all of which bear traces of extinct craters at their summits. At the extreme western end, again, the mountains rise to nearly 3000 feet. The coast is steep and rocky, and in some places the cliffs are 1400 feet high.

The first discovery of the islands is involved in uncertainty. Edrisi and Ibn al Vardi, the Arabian geographers, mention that beyond the Canaries to the northward are nine other islands in the western ocean. That these were the Azores seems more than probable, since their number is nine; special mention is also made of the abundance of a large species of Hawk, from which latter circumstance the Portuguese afterwards gave the name of "Azores," or Hawk-Islands. These writers further mention that they were very populous, and contained large cities, but that the population had been greatly reduced by intestine warfare. At the time of their European discovery they were uninhabited and covered with dense forests of underwood, and no trace of any former population existed.

In the year 1439 a Flemish merchant named Van der Berg, in sailing from Lisbon, is reported to have been driven on the shores of these islands. As soon as this intelligence reached the coast of Portugal, considerable interest was excited, and an expedition for further discovery was sent out under the navigator Cabral. In 1459 the islands began to be colonized and planted; and in consequence of the mild climate and the fertility of the soil, the population rapidly increased and cleared the land for cultivation. At the present time but little of the primeval forest remains, even in the remote districts; and where it does, it is much cut about by the inhabitants for fuel. The general character of the vegetation is unmistakably European; and though several plants from a more southern climate have been introduced, and have now become naturalized, thus giving it a more tropical aspect, there is no difficulty in pointing out these introductions.

The oranges of St. Michael's form the principal export trade of the Azores, whence vast quantities of this delicious fruit are annually sent to England. In the height of the season some eighty or ninety English schooners may be seen at one time anchored in the roadstead waiting for cargoes. They are most of them excellent sailing vessels and well manned; and much rivalry exists between the owners in making quick passages. Unfortunately there is no good natural harbour, and with a southerly or south-westerly gale vessels cannot remain in the roadstead, but are obliged to put out to sea and cruise about till the wind moderates. To supply this defect the inhabitants have commenced building a small harbour at Ponta Delgada; but such is the violence of the sea in winter that much of the work done the previous summer is carried away, and thus the completion delayed, as well as the expenditure considerably increased.

My brother stayed but a short time in St. Michael's, being obliged to return to England to rejoin his regiment. I remained there rather more than a month, during which time I visited different parts of the island, and collected all the specimens I could. My headquarters were at Ponta Delgada, where there is a very comfortable hotel kept by a most obliging English landlady. I also spent ten days at the Furnas, at the eastern end of the island, whence I had a better opportunity of exploring the mountains and wooded districts.

One of the prettiest and most interesting excursions to be made from Ponta Delgada is that to the Sete Cidades. The road, which for some distance is very good, passes between orange-gardens, fields of Indian corn and other crops, surrounded by high walls made of lava and volcanic stones, to serve as a protection from the wind. Near the village of Feteiras the road begins to ascend the mountain, and becomes narrow and bad, and must either be passed on foot or on donkeys. Here, too, the cultivated land ceases; and the rest of the way, to the top of the mountain, the track passes through stunted evergreens and tree-heath, which, from their exposed position, only attain a height of two or three feet, gradually getting smaller the higher we ascended. On reaching the top one suddenly finds one's self on the brink of a huge crater, nearly three miles in diameter, and a most beautiful scene bursts upon the view. Some thousand feet below lie two large and azure-blue lakes; and within the large crater are seven smaller ones, some of which are exceedingly perfect in form, and appear to have been the expiring efforts of the subterranean fire. On the borders of one of the lakes is a village of straggling white houses, which contrast pleasingly with the green of the general landscape, and afford a measure for the eye, showing the depth it lay below. The road to this village is cut out of the inner wall of the crater, and descends by a series of zig-zags, more agreeable to walk than to ride down. One of the smaller craters contains within its area a dense growth of native underwood, which has been left almost untouched by the inhabitants, on account of the difficulty of access. Here I thought I should be sure to find some new and peculiar insects amongst the decaying wood and fallen leaves; so, with some trouble, I scrambled to the edge and then let myself down the nearly perpendicular side, holding on by the branches of the trees and coarse grass till I reached the bottom. Having searched for a couple of hours, and not being rewarded by the discovery of any thing new, I left to visit the lake. At some time or other these lakes have been stocked with Goldfish (*Cyprinus auratus*), which have so increased in number that they literally swarm, and I caught some in my hands without difficulty. There appears to be so little food for such numbers (the water being nearly destitute of aquatic plants and insects) that I saw hundreds at one time dead upon the shore, which, from their emaciated state, seem to have been starved. As far as I am aware, there are no native fish in these lakes or in any of the freshwater streams. I obtained specimens of an Eel, and heard of the existence of a Pike; this latter, however, I never procured; but both are said to be recent introductions, and so are of but little interest.

Perhaps a more striking spot, but not so picturesque,

is the valley of the Furnas, with its Caldeiras or hot springs, at the eastern end of St. Michael's. This valley, too, is the ancient basin of a former gigantic crater. Its area is by no means level, and it contains large lakes and cultivated land; but the hot springs are its most interesting feature. About half a mile from the springs is a village, which boasts of a large hotel and several lodging-houses; these, during the season, are filled with visitors from various parts of the island, who come to drink and to bathe in the mineral waters. The largest of these springs is a circular pool, about twelve feet in diameter; in this the water, which is of a muddy appearance, boils up to a height of three or four feet, and then, running over, is conducted to some rough sheds, which are fitted up as baths, and are considered very beneficial in rheumatic and various other complaints. Not far from this is another, nearly circular spring, where the water also boils up as in a caldron, and escapes into a stream below. The water of this is clear, but contains large quantities of sulphur, silex, and other minerals, which are deposited round the edge, and incrust the rocks and sticks it meets in its course. I was told that this crust gradually grows over the top of the spring, so as to close up the orifice, when the water, from the pressure below, is thrown up to a height of ten or twelve feet into the air. An explosion then takes place, the obstruction is blown away, and the outlet again enlarged. All around this caldeira the earth is hot for a considerable distance, and incrusted with efflorescent sulphur and crystals of alum; and from every crack near issues steam and a strong smell of sulphuretted hydrogen. There are various other springs in the neighbourhood, some of which are cold. One, called the "Agoa azêda," much resembles soda-water, and comes sparkling out of the rock, and is not unpleasant to drink: others contain large quantities of iron, and cover the ground over which they flow with a rusty-yellow deposit. It is in such places that the inhabitants cultivate the *Caladium esculentum,* using it as a vegetable, and calling it "yam." The earth, warmed by the subterranean heat, supplies the tropical climate to which this plants belongs. Whilst I was at the Furnas I experienced a slight shock of an earthquake, enough to shake the house and all the things in the room, and remind me that I was living at the bottom of an old crater, the hot-springs in which still showed that the internal fire was not quite extinguished. This was the only one I felt while I was in the Azores, though slight shocks are said to be tolerably frequent.

I cannot pass on to the other islands without mentioning the gardens in the neighbourhood of Ponta Delgada, which surpass any I have ever seen elsewhere. The climate is so mild that many plants, which with us succeed only in a greenhouse or hothouse, here reach a degree of luxuriance hardly to be imagined. Except on the higher mountains, frost is unknown, even in winter; and Camellias and Azaleas, of the choicest varieties, may be frequently seen growing to a great height and covered with a perfect mass of blossom. Preeminent amongst the gardens, however, is that of Don José do Canto, who has been at considerable expense and trouble in importing plants from various parts of the world, many of which he grows with great success, as may be seen by an inspection of his grounds; for he is always most pleased to show any strangers over them who may take an interest in horticulture.

From St. Michael's we proceeded to Fayal, and thence visited the two most western islands. We left on the evening of the 21st of April, and at 6 o'clock next morning anchored in the harbour of Angra, the capital of Terceira. The whole day was spent on shore, as the steamer did not start again till night. This was the first really warm day we had experienced; and we made a long excursion into the interior. The scenery is not nearly so picturesque as that of St. Michael's, which surpasses any thing of the kind I have seen elsewhere. We returned on board in the evening; and the steamer again started for Graciosa, which was reached at 6 o'clock next morning. This island is one of the smallest of the Azores, and is very pretty. It appears to be highly cultivated; but as the vessel only remained there about an hour, and anchored at a considerable distance from the shore, we did not land. We next proceeded to St. George's, and, passing close under the high cliffs at the western end, called at Villa das Velas, on the south side, to land two or three passengers and a small amount of cargo.

Whilst coasting along the island, large flocks of Gulls, Puffins, and Rock-Pigeons were observed, evidently congregating for the breeding-season; and, judging from the rugged precipitous rocks they had chosen for this purpose, I fancy they ran little risk of being disturbed. The southern side of St. George's rises almost perpendicularly from the sea; but the inhabitants have found, nevertheless, sufficient space on some of the ledges of the rocks to plant vineyards. To many of these there is no approach on the land side, and the only means of getting to them is by landing from a boat, which of course can only be done in fine weather and with a calm sea, and climbing on hands and knees up the almost perpendicular cliffs.

The distance from Villa das Velas to Horta, the capital of Fayal, is about forty miles; but as we did not arrive at the latter place till past 11 o'clock at night, we remained on board till next morning. Fayal is a charming little island; and the view from it of Pico, with its snowy top peeping out from the clouds (as we then occasionally saw it), is really grand. On going ashore we found a comfortable boarding-house, kept by an American, where we put up during our stay in this island. There were also some American ladies from Boston, who had passed the winter here for the sake of the mild climate, and the captains of one or two American vessels which were in the harbour, who preferred living on shore to remaining on board their own vessels. I here made the acquaintance of Mr. Dabney, the U.S. Consul, who most kindly aided me in carrying out my natural-history pursuits. It was through him that in a few days I obtained a passage in a whaling-vessel to Flores, of which I gladly availed myself, the communication being very uncertain and irregular; in fact, during the previous winter, Corvo and Flores had been without communication with any other place for five months.

Mr. Brewer and I went on board the barque 'Henry Tarbert;' and having a favourable breeze, in sixteen hours we sighted the south point of Flores; but on approaching the shore too heavy a surf was running to allow us to land; so we were obliged to cruise off and on for three days more, till the swell had in some measure subsided. During this time we were continually in hopes of seeing a whale captured, a constant look-out at the mast-head being kept. Unfortunately, none were sighted; so, when the opportunity offered, we were not sorry to take leave of the whaler, though the captain, who was an American, had been most kind to us throughout. The sea was calmer, but a heavy surf still beat upon the rocky shore, when the captain ordered one of the whale-boats to be lowered and manned. We were shipped off with our luggage, and rowed as near the rocks as was prudent, care being taken to keep just outside the breakers. Numbers of people working in the fields and elsewhere had been watching us; and as soon as they perceived that we intended to land, they flocked down to the shore and completely lined the rocks. There were probably not less than three or four hundred persons. They beckoned to us to run our boat in behind a certain rock which they pointed out; but as the sea broke heavily upon it, and appeared to dash all over it, our captain would not make the venture with his slight whale-boat, and signalled to them to send off another boat to show the way. This, after some delay, they did; then came a nasty landing. The boatmen, who certainly managed their craft very well, came out to us, and with some difficulty we exchanged boats, the native one being heavier and stronger. Then, watching their opportunity, and following as close as possible the wake of a huge breaker, they shot the boat round the corner of the rock into a small creek, which was a little sheltered from the full violence of the surf. We now had to jump out, which would have been no easy matter without the assistance of those on the rocks, who, as the swell of the waves raised our boat some 8 or 10 feet, made a snatch at us, and dragged us safely on shore. The moment we landed we were surrounded by a crowd, so that we could scarcely move—some, from mere curiosity, staring at us with their mouths open in a vacant way, others asking every conceivable question as to our business &c., and for any news that might have occurred during their five months' imprisonment. At last we were glad to escape into the house of a person who called himself Consul, though to what nation I have no idea; and there, forcibly keeping the crowd out, we made arrangements to go to Santa Cruz. We had landed at Largens, a small town at the southern point of the island. From the vessel it appeared a sheltered spot; but it was several miles distant from Santa Cruz, the village we wanted to go to, and where Mr. Mackay, H.B.M. Consul, lived, to whom we had letters of introduction. My first idea was, that, once safely on *terra firma*, I would have no more boating; but this notion I was soon obliged to give up; for on inquiry I found the way by land was very bad and circuitous, and that no beasts could be procured. More than this, the track lay in many places along the ledge of a cliff overhanging the sea; and I felt sure that my companion, who had a bad head for such places, would be unable to pass, even if I could do so.

After not a little bargaining, we agreed with some boatmen to row us to Santa Cruz in a large eight-oared boat, the distance by water being about sixteen miles. Having transferred all our baggage to our new craft, and wished our late captain farewell, we started, keeping as near the shore as the breakers would allow. Sometimes we were on the crest of a wave within a few feet of where it was actually breaking, and almost the moment we were over it the whole line where we had passed broke with a fearful roar behind us. Had the boatmen misjudged their distance and kept a little too near the shore, nothing could have saved us. They certainly understood what they were about; but the way they set to work was not such as to inspire a stranger with confidence. At the approach of a larger and more threatening wave than usual, all talked at once, and each gave advice as to what ought to be done, instead of obeying the master; and they seemed as ready to listen to a boy of twelve years of age as to an old and experienced seaman.

We reached Santa Cruz in about four hours; but it was at first uncertain whether we could land there. However, some of the crew thought we could; and after watching our opportunity much as we did at Largens, the boat was run ashore behind the rocks. We proceeded at once to Dr.

Mackay's house, accompanied by a large number of the inhabitants of the town; and he soon found us a lodging. Dr. Mackay is an English medical man, who settled here some years ago, and for some time was H.B.M. Consul; his son, Mr. Mackay, now holds this post, though there is not much traffic for English vessels. The latter told me that he had never left the islands but on one occasion, and then against his will. In his capacity as Consul, he went on board a vessel bound for England, which was calling at Flores for fresh provisions. While below with the Captain it came on to blow hard, and the vessel was unable to remain where she was, the anchorage being bad; and they were afraid of her drifting ashore. She therefore stood out to sea for safety. As the wind increased, there was no chance of a small boat being able to go ashore; and they were short of provisions; so the Captain sailed for England, and after a few days passage landed Mr. Mackay on his native shore, to his infinite disgust.

Flores is much better watered than most of the islands, and the climate appears to be more rainy than that which prevails around the eastern groups; but little of the land is cultivated. In the mountains are several lakes, and a large swamp nearly two miles long by a mile broad, where I found Ducks, Teal, and Snipes breeding, though not in great numbers. Woodcocks also are abundant, as they are but little disturbed; in fact scarcely any one here shoots. I remained in Flores rather more than a fortnight, during which time I walked all round the coast; and I also made several excursions to the lakes and swamps in the mountains, thinking that the uncultivated parts of this thinly inhabited island would probably yield more native treasures than I had hitherto found. With the exception, however, of the Purple Sandpiper and the Wheatear, and some beetles I procured from some rotten wood of the tree Euphorbia (*Euphorbia mellifera*), I found nothing specially different; and I have no doubt that the number of species both of animals and vegetables is fewer than in the eastern islands. The Euphorbia I never found growing in any quantity; and where it does occur it is only an isolated plant in some deep damp ravine. I looked out specially for it, as I was aware that Mr. Wollaston had found several very new and interesting beetles in these plants both in Madeira and the Canaries, where they form a peculiar feature in the flora. On the three or four occasions when I came across plants having dead or rotten sticks I gathered them carefully into a bundle, wrapped them in a cloth, and carried them home, so that I might make a more careful examination of them, and that nothing they contained, however small, might escape my notice. On breaking these up, curiously enough, I found them full of insect-life, but composed of very few species, mostly very minute, but allied to some of the Euphorbian species of Madeira. I have little doubt that this plant was more common formerly, before the lowlands were so much cleared for cultivation; at present, however, it is very scarce.

The weather was still rough and uncertain; and as it was scarcely safe to visit Corvo in a small boat, though but sixteen miles distant from Santa Cruz, I chartered a schooner of about thirty tons for this purpose, and spent two days on the island. It is a single volcano, containing a large crater with lakes, and rising precipitously from the sea. It is surrounded by cliffs, except in one spot on the south, where both the village and landing-place are situated. It is but seldom visited, and presents little of special interest to the traveller who has visited the other islands. I returned again to Flores, and left a day or two after in the same schooner for Fayal. Here I remained some days, waiting for a favourable opportunity to ascend Pico; but though I made the attempt, the weather, unfortunately, prevented my succeeding; and with some regret at not being able to accomplish this object I finally left for Terceira, calling at St. George's on my way. Pico, at the base, is more thickly wooded than most of the other islands. The volcano rises to 7600 feet in height; and in winter the extreme cone is frequently covered with a thin layer of snow and is destitute of vegetation, with the exception of a few lichens; descending lower is a belt of coarse grass and the tree heath; while lower again and extending to the sea is a dense growth of brushwood, consisting of laurustinus, faya, and other evergreen trees. This is the home of the Wood-Pigeon in the Azores. In most of the other islands this bird is much persecuted, on account of its being greatly esteemed for the table, and it is consequently very scarce and wild; here, however, it is abundant, being comparatively undisturbed. Towards the coast, at the foot of the mountain, a network of walls encloses small vineyards to protect them against the winds. From these vineyards was formerly made the well-known "Pico Madeira;" but for the twelve years previous to my visit they had yielded no fruit, in consequence of the vine-disease, causing a great loss to the poor inhabitants.

After leaving Fayal I spent a fortnight more at Terceira, and returned by steamer to Lisbon, staying three days at St. Michael's on my way. My own time being limited, I was compelled to return to England. So I sent Mr. Brewer to visit St. Mary's, and make collections for me; but the summer having now fairly set in and the vegetation being much burned up by the heat, insects, which formed the principal object of his excursion, were very scarce. I returned direct to England; and he followed me about a month later.

MAMMALIA.

There are but seven mammals found wild in the Azores; and of these six have undoubtedly been introduced, either intentionally or accidentally, by man. Some of them do not seem, as yet, to have established a footing beyond the eastern group of islands. They are all of them common European species, and therefore do not throw much light upon the subject as to whence the fauna is derived. The following is the list:—

1. Lepus cuniculus, Linn.

Lepus cuniculus, Drouet, Faun. Açor. p. 108.

Hab. St. Michael's; Europe; Madeira.

Found wild in St. Michael's, where it frequents the woods, especially in the neighbourhood of the Furnas. It is nowhere very abundant, being rarely found in the low and cultivated parts. From a careful examination of two specimens I brought home, I do not find a similar variation in colour to that which Mr. Darwin mentions as observable in the Rabbits of Porto Santo when compared with European specimens. Azorean are identical with English individuals, except that they are rather smaller than the average. This difference is no more than we find in England when specimens from barren heaths where food is scarce are selected for comparison.

2. Mustela furo, Linn.

Mustela furo, Drouet, Faun. Açor. p. 106.

Hab. St. Michael's and St. George's.

I never met with this species myself, but was told by a Portuguese gentleman that it has established itself in the mountains in the neighbourhood of the Furnas, where it feeds principally on the Rabbits.

3. Mustela vulgaris, Linn.

Mustela vulgaris, Drouet, Faun. Açor. p. 106.

Hab. St. Michael's; Terceira; Fayal; Europe.

In all the three above-named islands this species is found; in none, however, is it common. It has no doubt followed the rats on board the vessels. I am not aware that it has become established in any of the other islands.

4. Mus decumanus, Pall.

Mus decumanus, Drouet, Faun. Açor. p. 107.

Hab. Cosmopolitan.

Common throughout the islands.

5. Mus rattus, Linn.

Mus rattus, Drouet, Faun. Açor. p. 107.

Hab. Azores; Europe.

Frequents chiefly the gardens and orange-plantations, where it does considerable mischief by climbing the trees and eating out the pulp of the fruit. This species is still common in many parts of Portugal.

6. Mus musculus, Linn.

Mus musculus, Drouet, Faun. Açor. p. 107.

Hab. Cosmopolitan.

These six species must be considered established; some of them, however, always follow close upon the footsteps of man.

7. Vesperugo leisleri, Kuhl.

Vesperugo leisleri, Drouet, Faun. Açor. p. 105.

Hab. Azores; Europe.

This common European species may frequently be seen flying about during the daytime. Towards night they are very numerous in the neighbourhood of towns and gardens. It is the only really native mammal in the Azores. For its determination I am indebted to the kindness of Dr. W. Peters of Berlin, to whom I sent the specimens I collected.

AVES *.

MM. Morelet † and Drouet ‡ are the only naturalists, so far as I am aware, who have visited the Azores and written upon their ornithology; but their *spécialité* being conchology, the birds did not receive so much attention as they deserved. Morelet, however, gives an enumeration of thirty species, and Drouet of forty-six, which they consider belong to the islands. My own experience leads me to modify these lists, as some birds contained in them are certainly stragglers, whilst other residents must be added, and, again, no less than ten species are included in Drouet's list which, as he says, exist only in a domesticated state, and therefore bear no reference to the natural fauna. These ten species are as follows:—*Fringilla canaria, Columba livia*, var. *domestica, Phasianus colchicus, Meleagris gallopavo, Numida meleagris, Pavo cristatus, Gallus domesticus, Anser ferus, Anser* ——? §, *Cygnus atratus*. These I have thought it best to omit.

On reference to M. Pucheran's paper in the 'Revue et Magasin de Zoologie,' for 1859, p. 409, it would appear that these two gentlemen obtained only four specimens of birds from these islands, whence I infer that the rest of their lists were formed from observations casually made

* This article is a reprint, with alterations, of my paper "on the Birds of the Azores," published in 'The Ibis' for 1866, p. 88.

† Notice sur l'Histoire Naturelle des Açores, par A. Morelet. Paris, 1860.

‡ Eléments de la faune Açoréenne, par H. Drouet. Paris, 1861.

§ *Anser* ——?, or Pato real, as they call it in the Azores, = *Carina moschata*, which, of course, is only found in a domesticated state.

and not corroborated by the collection and critical examination of specimens. That my own catalogue includes all the stragglers I do not pretend to say; on the contrary, I have no doubt that the number may be considerably increased. With regard to the residents I believe it will be found tolerably complete. I either obtained or examined specimens of each species mentioned in my list, with the single exception of *Picus minor*, which did not come under my own observation, but is included on the authority of Mr. Brewer. Scarcely a storm occurs in spring or autumn without bringing one or more species foreign to the islands; and I have been frequently told that Swallows, Larks, Grebes, and other species not referred to here are not uncommonly seen at those seasons of the year.

A glance at the following list will show at once its entirely European stamp. Every species, except *Thalassidroma wilsoni*, an oceanic wanderer of the North-western Atlantic, is to be found in Europe, or in the outlying provinces of the European fauna (North Africa, the Madeiras, and Canaries). From this generalization two more exceptions must be made:—one in the case of the Chaffinch, which is identical with the *Fringilla tintillon* of the Madeiras and Canaries; and the Bullfinch, to which *Pyrrhula europæa* or *P. coccinea* must be considered most nearly affined. This last species seems peculiar to the group. As regards the local peculiarities of Azorean birds, there is certainly a tendency among them to vary, more or less, from their continental representatives. This is especially shown by the former always having darker plumage and stronger bills and legs. In some cases the variation is not greater than may bé observed in extreme examples from a large series of continental specimens of the same species; in others it becomes more remarkable; and in *Fringilla tintillon* and *Pyrrhula murina* the deviation is carried to such an extent that it is impossible to speak of them but as good species.

The list further shows the gradual falling off in the number of species inhabiting each group of the archipelago as we proceed westward and away from the Old World. Before making this comparison, it seems necessary to take into consideration what species should properly be included. I think that when we find birds having in most places habits so essentially migratory as the Quail, Woodcock, and Snipe, here becoming resident throughout the year, and losing their wandering instincts from the necessity of their situation, we may fairly except from our calculation the Gulls, Terns, and Petrels, for which these islands simply afford a resting-place in their wanderings, and a resort during the breeding-season. All others would appear, I think, to have arrived involuntarily, having been blown over by storms, or through some other such agency. The eastern group has forty species, the central thirty-six, and the western twenty-nine; so that we have a gradual diminution of the number of species as we proceed westward from the Palæarctic fauna. This seems clearly to show that storms or other external causes have been the means of peopling these islands with bird-life. That the nearest group has caught the most stragglers must be admitted; and that storms do bring stragglers, the occurrence, as afterwards mentioned, of such birds as the Snow-Bunting and Golden Oriole shows. Extending this observation so as to include the other Atlantic groups of islands, we find that the number of species in each group varies inversely as the distance of the group from the continent. Thus the number of species recorded by Vernon Harcourt * as found in Madeira is ninety-five; the Canaries have one hundred and six †; while here, in the

* Ann. & Mag. Nat. Hist. 1855, p. 430.

† Webb and Berth. Orn. Can.

Azores, the most distant group, only fifty-one occur. It seems tolerably certain that, were it not for the constant persecution carried on by the inhabitants, many species, arriving in sufficient numbers, would be able to establish themselves as permanent residents; and a few years would make, from this source alone, some accession to the legitimate avifauna of these islands. Hundreds of Serins are caught for cages; and the Red-legged Partridge has been exterminated for the table in St. Michael's; it may therefore be not unjustly inferred that other species have been affected in like manner.

The following is a list of birds collected or observed during my visit: those marked with a dagger (†) I believe to be stragglers; the rest are residents.

1. †TINNUNCULUS ALAUDARIUS, G. R. Gray.

Hab. Azores, eastern group (St. Michael's); Africa; Europe; Madeira; Canaries.

I obtained a single specimen of this species through Mr. G. Brown of St. Michael's, who kindly preserved it for me. It is not a resident in the Azores, though in Madeira Mr. Vernon Harcourt says it is common *.

2. BUTEO VULGARIS, Bechstein. "Milhavre."

Buteo vulgaris, Drouet, Faun. Açor. p. 114; Morel. Hist. Nat. des Açor. p. 83.

Hab. Azores, eastern and central groups; Europe; Madeira; Canaries.

Very common in the eastern and central groups, but scarcely ever seen in Flores or Corvo. I found a pair building in a cliff near the sea on the 22nd of March, in St. Michael's, and shot both the old birds. Mr. Gurney,

* See "Notes on the Ornithology of Madeira" in Annals and Magazine of Natural History, 2nd ser. vol. xv. pp. 430–438.

who has kindly examined them for me, says that they are unusually rufous on the under parts of the body, and that they consequently bear a considerable resemblance to the immature dress of the ordinary Buzzard of Barbary, *Buteo desertorum* (Daudin). In size, however, they agree with *B. vulgaris*, which is rather the larger bird of the two. In the Azores Buzzards are by no means shy, and may constantly be seen hovering over the towns or perching in the orange-gardens. They feed chiefly on young rabbits, rats, and mice, of which there is a great abundance. It is from this bird the islands take the name of Azores *.

3. Asio otus (Linnæus).

Hab. Azores, eastern and central groups; Europe; Canaries.

Only a single example of this species came under my notice. Mr. Dabney procured it in Fayal during my absence in Flores, and kindly had it preserved for me. It was a very young bird, and was brought to him by a boy who took it from the nest. I had frequently heard of it in St. Michael's; but it is nowhere common, and I never met with it living. As far as I could learn, this bird is not distinguished by the inhabitants of the Azores by any special name. It does not appear to be generally known as a resident.

4. Strix flammea, Linnæus. "Coruja."

Strix flammea, Morel. Hist. Nat. des Açor. p. 83; Drouet, Faun. Açor. p. 115.

Hab. Azores, eastern and central groups; Europe; cosmopolitan; Madeira; Canaries.

Occasionally met with in the eastern and central groups.

* Açor, in Portuguese, is properly the Kite (*Milvus ictinus*), for which species no doubt the early explorers mistook this bird.

In Flores and Corvo I did not find any one who either knew the bird or the Portuguese name for it; hence I conclude it does not extend to these outer islands. Several people in St. Michael's and Terceira told me they had seen it; but I was unable to procure a specimen. The captain of a whaling-vessel told me that one flew on board his ship when about 500 miles S.W. of the Azores. It was much exhausted, but he kept it alive on salt pork for three or four days.

5. Turdus merula, Linnæus. "Melro."

Turdus merula, Morel. Hist. Nat. des Açor. p. 84; Drouet, Faun. Açor. p. 119.

Hab. Eastern, central, and western groups; Europe; North Africa; Madeira; Canaries.

Frequents the mountain-districts rather than the gardens and low country. It is very common, but shy. The note always struck me as harsher and louder than our Blackbird's; but I find no difference in the size or form of the two birds.

6. †Oriolus galbula, Linnæus.

Hab. Azores, western group; Europe; North Africa; Madeira.

Whilst I was in Flores a bird I believe to have been of this species was caught and killed by some boys, who plucked all its feathers out and threw it away. I did not hear of it till the following day, when I went immediately to the village, but could procure no more than its tail- and wing-feathers, from which, together with the description I received, I do not hesitate to attribute it to this species.

7. Erythacus rubecula (Linnæus). "Avinagreira."

Motacilla rubecula, Morel. Hist. Nat. des Açor. p. 84.

Erythacus rubecula, Drouet, Faun. Açor. p. 120.

Hab. Azores, eastern and central groups; Europe; North Africa; Madeira.

Though common in the eastern and central groups, the Redbreast does not occur in the two western islands. Through the kindness of some of my friends I have been able to compare my specimens with examples from Algeria, Tunis, and Southern Italy, with which I find that they exactly agree in their light-coloured plumage. Mr. Gould showed me one he shot in Teneriffe, which is precisely similar to our British and darker form.

8. SYLVIA ATRICAPILLA (Linnæus). "Toutinegro."

Sylvia atricapilla, Morel. Hist. Nat. des Açor. p. 84; Drouet, Faun. Açor. p. 119.

Hab. Azores, eastern, central, and western groups; Europe; Madeira; Canaries.

Abundant in the lower lands throughout the islands. A curious variety is not unfrequently met with, having the black marking on the head extending to the shoulders and round under the throat. I only saw one individual, which was in a cage with a common Blackcap. It appeared to be slightly larger, though in other respects the same, with the exception, of course, of the dark markings. I was told that some individuals have the whole of the under parts of the body black. The story current in the Azores with regard to them is, that, when the parent lays more than four eggs, one bird always proves to be this variety *. In Fayal it is known by the name "Avinagreira," a term given to the Redbreast in St. Michael's. It is much prized

* This variety is doubtless the same as that mentioned by Heineken (Zool. Journ. v. pp. 75–79) as occurring in Madeira, where a similar story is also told concerning its origin. It was subsequently described by Sir W. Jardine (Edinb. Journ. Nat. & Geog. Science, Jan. 1830, i. p. 243), and figured by him and Mr. Selby in their 'Illustrations of Ornithology,' pl. 94, under the name of *Curruca heinekeni*.

by the Portuguese, who are fond of keeping it as a cage-bird.

9. Regulus cristatus (Linnæus). "Estrellinha."

Regulus cristatus, Morel. Hist. Nat. des Açor. p. 84; Drouet, Faun. Açor. p. 119.

Hab. Eastern, central, and western groups; Europe; Eastern Asia.

Frequents chiefly the Junipers (*Juniperus oxycedrus*) and Tree heaths (*Erica azorica*) in the mountains, and is but seldom seen in the gardens or lower country. I have compared my examples with British and South-European specimens, and find that the former are rather stouter and stronger in the beak and legs, and also somewhat longer in the tail. Mr. Gould showed me a Golden-crested Wren from Eastern Asia which agrees with my Azorean bird in all respects. In Madeira this bird is represented by an allied species, *R. madeirensis*, V. Harc.

10. Saxicola œnanthe (Linnæus).

Hab. Azores, western group; Europe; N. Africa; Iceland; Greenland; Nova Scotia and Labrador.

I shot a single example of the Wheatear in Flores, after a strong gale of wind, and I at first believed it was a straggler from the continent; but I afterwards found four or five pairs in the old crater on Corvo, which had bred there, as I saw youug birds that could scarcely fly. The inhabitants have no name for this bird, and I did not meet with any one who knew it; so I believe it to be a recent settler.

11. Motacilla sulphurea, Bechstein. "Alveola" or "Lavandiera."

Motacilla boarula, Morel. Hist. Nat. des Açor. p. 84; Drouet, Faun. Açor. p. 120.

Hab. Azores, eastern, central, and western groups; Europe; Madeira; Asia.

Common, wherever there is water, throughout all the islands. I have compared it with European specimens, with which it agrees well, with the exception of the tail being rather shorter. Mr. Gould, however, showed me some examples from Eastern Asia which in this respect are exactly the same as the Azorean bird. It is resident the whole year.

12. †Plectrophanes nivalis (Linnæus).

Hab. Azores, western group; N. Europe; Canaries; Iceland; Greenland; N. America.

A flock of about twenty of these birds appeared during the winter of 1864–65 in the island of Corvo. They were said to have been much exhausted when they arrived, and several were caught and kept in cages. At the time I was there I believe there was but one living, and this was a female. The owner had such an exalted notion of its value that I did not procure it. After I returned to England, Mr. J. P. Dabney kindly sent me a skin of a bird of this species which was killed in Fayal.

13. Fringilla tintillon, Webb et Berth. Orn. Canar.; Bocage, J. de Sci. Math. Lisb. i. p. 89.

Fringilla canariensis, Vieill. Nouv. Dict. d'Hist. Nat. xii. p. 232, et Enc. Méth. p. 953.

Fringilla moreleti, Puch. R. Z. 1859, p. 413; Morel. Hist. Nat. des Açor. p. 84; Godm. Ibis, 1866, p. 97.

Fringilla canariensis, var. *moreleti,* Drouet, Faun. Açor. p. 117.

Hab. Azores; Madeira; Canaries.

In order to satisfy myself as to the validity of the species described by Dr. Pucheran under the name *Fringilla moreleti,* and to ascertain for certain whether it really differs from *F. tintillon*, the species common to the other Atlantic groups, I availed myself of a recent opportunity to compare the specimens I collected in the Azores with the types of *F. tintillon* and *F. moreleti* in the Museum of the Jardin des Plantes in Paris. The result is, I do not hesitate to say that there is but one species of Chaffinch in the Atlantic islands, which is shared in common by the Azores, Madeira, and Canaries. A close examination of a large series of specimens shows that considerable variation in size, and some in colour, exists without reference to locality. Out of thirty specimens from the Azores some have the bill larger than others; in some the green gloss of the back begins at the nape of the neck and spreads over the whole upper surface to the tail, in others this colour is very slightly shown, and one female is destitute of any such colouring at all. Three Madeiran specimens have the lateral tail-feathers nearly white, a fourth has very little white on the tail. The same variation is shown in the Azorean specimens. Some males have the frontal feathers black; others, again, are without this mark.

My specimens from Madeira and the Azores were all collected between the middle of April and the end of June, and are in breeding-plumage. I do not know at what season of the year the Teneriffe specimens were procured; but all three are males. The two specimens in Paris, as well as the one lent me by Prof. A. Newton, all have the tarsus lighter-coloured than the Azorean birds; but the latter having been in spirits, and the former exposed for years in the gallery of the Museum, I do not attach any importance to this apparent difference.

Being, then, unable to reduce the variations observable in the Chaffinch of the Atlantic islands to any sort of law, I have no alternative but to consider, with Prof. Barboza du Bocage, that there is but one species, and that *Fringilla moreleti* must be considered a synonym of the older title *F. tintillon.* The name which should stand, by the strict law of priority, is *F. canariensis,* Vieill.; but as this name is so liable to be confounded with *F. canaria,* Linnæus, I think Webb & Berthelot's appellation had best be adhered to.

14. PYRRHULA MURINA. "Priolo" and "Prior."

Pyrrhula coccinea, Morel. Hist. Nat. des Açor. p. 84; Drouet, Faun. Açor. p. 115.

Hab. Azores, eastern group.

M. Morelet brought away but one specimen of this bird from the Azores; and M. Pucheran referred it to *P. coccinea* of De Sélys-Longchamps*, the female of which it somewhat resembles. The male, however, differs materially from the same sex of that species, not having the red breast or white rump, which last character is also shared by the female. In a recent notice of my paper in the 'Ibis' for 1865, by M. Barboza du Bocage, published in the 'Jornal de Sciencias Mathematicas' of Lisbon (vol. i. p. 89), considerable doubts are cast upon my observations respecting the determination of the sexes of this species. Let me here assure M. Bocage that I am perfectly satisfied that my dissections were correctly made. Out of nine birds examined just prior to the breeding-season, five were certainly males. Since my return to England *nineteen* additional specimens have been sent me, which, though not dissected, show no traces whatever of the red plumage so conspicuous in the European species. As M. Bocage adduces nothing but his own preconceived opinion in opposition to the facts stated in my description, I feel I can confidently await the advent of some more "competent person" who shall provide M. Bocage with the "indispens-

* *Cf.* 'Ibis,' 1859, p. 322; 1860, p. 93; *et* 1861, p. 401.

able elements" to enable him to determine autoptically this point about which he expresses so much incredulity.

In habits the Bullfinch of the Azores resembles our *P. vulgaris,* feeding principally upon insects and the buds of trees. I believe that it is confined to the mountainous parts of St. Michael's, where it is tolerably abundant. It is so tame that it takes but little notice of the report of a gun, and I shot thirteen individuals in the same poplar tree in a few minutes. I know nothing of its breeding-habits, as I was not in the island at the right time of year, and it is difficult to obtain accurate information from the inhabitants.

15. Serinus canarius. "Canario."

Fringilla canaria, Linn. Syst. Nat. i. p. 321.

Serinus canarius, Bolle, J. f. Orn. 1858, p. 128, t. 1.

Fringilla serinus, Morel. Hist. Nat. des Açor. p. 84; Drouet, Faun. Açor. p. 116.

Serinus hortulanus, Godm. Ibis, 1866, p. 98.

Hab. Azores, eastern, central, and western groups; Madeira; Canaries.

Found in abundance throughout the Azores. It frequents the cultivated lands, where it feeds on the seed-crops, and is especially destructive to the flax. From its well-known powers of song it is often caught and tamed, a great many being sold on board vessels which touch at the islands for provisions. In Fayal these birds congregate towards evening in considerable numbers about a small hill near Horta, and fly across in a body to the island of Pico; for what reason I do not know, as there is no want of trees in the neighbourhood. Azorean specimens of the Canary agree with others from the Canaries, so well described by Bolle. The species chiefly differs from the Serin in having the feathers of the back edged with ashy instead of yellow.

16. Sturnus vulgaris, Linnæus. "Esturninho."

Morel. Hist. Nat. des Açor. p. 84; Drouet, Faun. Açor. p. 119.

Hab. Azores, eastern, central, and western groups; Europe; Madeira.

Plentiful throughout all the islands of the archipelago. It breeds principally in the sea-cliffs, and is precisely similar to European examples. When vines were more cultivated it was much destroyed, as it was said to feed upon the grapes and to do much mischief in the vineyards; lately, however, it has not been persecuted, and has greatly increased in numbers.

17. Dryobates minor (Linnæus). "Picapoa."

Hab. Azores, eastern and central (?) groups; Europe.

This bird is very uncommon, but is occasionally met with in the mountains in St. Michael's, and, I believe, also in Terceira. I was unable to procure a specimen, and did not meet with it myself. Mr. Brewer tells me that after I left for England he saw one at the Furnas and watched it for some time, and has no doubt as to the species. MM. Morelet and Drouet give *Picus major* in their lists of the Birds of the Azores; but I am not aware that they ever obtained specimens, and I am inclined to think that there is but one resident species.

18. †Upupa epops, Linnæus.

Hab. Azores, eastern and central groups; Europe; N. Africa; Madeira; Canaries.

I saw a single example in a collection at Terceira. It had been killed in that island some years previously. It has also been met with in St. Michael's.

19. Columba palumbus, Linnæus. "Pomba troqual."

Columba trocaz, Morel. Hist. Nat. des Açor. p. 84; Drouet, Faun. Açor. p. 122.

Hab. Azores, eastern and central groups; Europe; N. Africa; Madeira.

Since I left the Azores, Mr. J. P. Dabney has most kindly sent me two skins of this bird, which, I believe, he procured at Pico. They arrived, unfortunately, in bad condition. The plumage appears to be slightly darker than in British examples, but in other respects I detect no difference. This bird is only found in the eastern and central groups, and is most common in St. George's and Pico. It is undoubtedly this species which is mentioned by MM. Morelet and Drouet under the name of *Pomba trocaz*. The true *P. trocaz* is an exceedingly fine and distinct species, darker in plumage, and wanting the white ring on the neck, and cannot be confounded with *C. palumbus*. I have two specimens of it from Madeira.

20. Columba livia, Linnæus. "Pomba da rocha."

Columba livia, Morel. Hist. Nat. Açor. p. 84.

C. turricola, Drouet, Faun. Açor. p. 121.

Hab. Azores, eastern, central, and western groups; N. Africa; Europe; Madeira; Canaries.

Exceedingly common throughout the archipelago. It breeds in great numbers in the rocky cliffs along the coast. Most of my examples are very dark in plumage—so much so, indeed, that the band on the wings is no longer visible. I saw, however, two quite white individuals, but not a single one of the pale grey tint usually found elsewhere. I find that Mr. Vernon Harcourt mentions a dark variety in Madeira.

21. Caccabis rufa (Linnæus). "Perdix."

Perdix rubra, Morel. Hist. Nat. des Açor. p. 84; Drouet, Faun. Açor. p. 123.

Hab. Azores, eastern and central groups; West Europe; Madeira.

I had unfortunately no opportunity of visiting St. Mary's, in the mountains of which the Red-legged Partridge is said to be very abundant. Mr. Brewer, who went there after I left, procured me two examples. It is occasionally found in St. Michael's and Terceira.

22. COTURNIX COMMUNIS, Bonnaterre. "Cordonix."

Perdix cothurnix, Morel. Hist. Nat. des Açor. p. 84.

P. coturnix, Drouet, Faun. Açor. p. 124.

Hab. Azores, eastern, central, and western groups; Africa; Europe; Asia; Madeira; Canaries.

Plentiful in the cultivated lands on all the islands, and even in the gardens. It is not migratory here, and is said to have two, and sometimes even three nests in the year. It is certainly exceedingly numerous, and affords excellent sport; on one occasion a Portuguese gentleman and I killed 157 in a few hours.

23. ÆGIALITES CANTIANUS (Latham). "Maçarico."

Hab. Azores, eastern and central groups; Old World; Madeira; Canaries.

I met with a few birds of this species about the lakes in St. Michael's; but afterwards found them more plentiful about Capellas, in Fayal, and on the high ground between Angra and Praya, in Terceira. The Portuguese name "Maçarico" or "Maçanico" is applied more or less to all Sandpipers and Snipes, as well as to this species. It breeds in Terceira, as I saw several young birds about, which were unable to fly.

24. †VANELLUS CRISTATUS, Meyer.

Hab. Azores, central group; N. Africa; Europe; Madeira; Canaries.

I saw a single stuffed specimen in the collection of a gentleman at Angra. He informed me that it had been shot in Terceira. Mr. Alfred Newton tells me he has a specimen from Madeira.

25. Strepsilas interpres (Linnæus).

Hab. Azores, eastern, central, and western groups; cosmopolitan.

A few pairs of Turnstone are always to be found about the rocks between Santa Cruz and Ponta Delgada, in Flores. I killed some specimens in June in full nuptial plumage, and I suspect that it must breed on some of the small islands near the coast; but the weather was so stormy all the time I was in Flores, that I was unable to get out to them. It is said to remain there the whole year. I afterwards saw eight birds of this species near Capellas, in Fayal, and I believe a few are to be found on the coast of any of the islands where the rocks provide sufficient protection from the surf.

26. Ardea cinerea, Linnæus. "Garça real."

Ardea cinerea, Drouet, Faun. Açor. p. 125.

Hab. Azores, eastern and central groups; Europe; Madeira; Canaries.

This is the only resident species of Heron. A few pairs are always to be met with about the lakes of St. Michael's, and occasionally on the coasts of the other islands; but the sea is in most places too deep for them to fish from the shore. I shot a single specimen in St. Michael's, at the Sete Cidades, and saw the remains of an immature bird that had been killed some time previously, which leads me to believe that the species occasionally breeds there, though I could obtain no information as to the fact.

27. †Ardea purpurea (Linnæus).

Ardea purpurea, Morel. Nat. Hist. des Açor. p. 84; Drouet, Faun. Açor. p. 125.

D

Hab. Azores, central group; North Africa; Europe; Madeira.

28. †Ardea alba (Linnæus).

Hab. Azores, central group; S.E. Europe.

29. †Ardea garzetta (Linnæus).

Hab. Azores, central group; S.E. Europe; Africa; Canaries.

30. Ardea egretta (Linnæus).

Hab. Azores, central group; S.E. Europe; North Africa.

I saw examples of these four species in a collection in Terceira. They were all said to have been killed in that island.

31. †Ardetta minuta (Linnæus).

Hab. Azores, central group; Europe; Madeira.

32. †Botaurus stellaris (Linnæus).

Hab. Azores, eastern and central groups; Europe; Madeira.

Also in the same collection. One of the latter species was killed in St. Michael's.

33. †Platalea leucorodia (Linnæus).

Hab. Azores, eastern group; N. Africa; Europe; Madeira; Canaries.

A Portuguese gentleman in St. Michael's told me that five or six examples of this species had been shot at Sete Cidades a few years previously.

34. Numenius arquatus (Linnæus). "Maçanico real."

Hab. Azores, eastern, central, and western groups; N. Africa; Europe; Madeira.

35. Numenius phæopus (Linnæus). "Maçanico real."

Hab. Azores, eastern, central, and western groups; Africa; Europe; Madeira; Canaries.

These two species are occasionally found about the coasts. I saw them both, but I much doubt their breeding there regularly.

36. Scolopax rusticola, Linnæus. "Galinhola."

Scolopax rusticola, Morel. Hist. Nat. des Açor. p. 84; Drouet, Faun. Açor. p. 125.

Hab. Azores, eastern, central, and western groups; Europe; Madeira; Canaries.

In all the mountain-districts throughout the islands this bird is not uncommon. It breeds, as with us, early in March, as I found young birds in the beginning of April, whilst out rabbit-shooting. The native sportsmen (!) shoot them while flying of an evening. It is most abundant in St. George's, Pico, and Flores, where few people kill them.

37. Gallinago media, Leach. "Maçanico real."

Gallinago media, Morel. Hist. Nat. des Açor. p. 84; Drouet, Faun. Açor. p. 125.

Hab. Azores, eastern, central, and western groups; Europe; Asia; Africa; Madeira; Canaries.

A few Snipe are occasionally found about the streams and wet places in the mountains. In Flores I saw four or five pairs on a large marsh, where I have no doubt they were breeding, though I did not see a nest. This species is called by the same name as the Curlew and Whimbrel.

38. Tringa maritima, Brünnich.

Hab. Azores, western group; N. Europe; Greenland; N. America.

A small flock was usually to be seen in company with some Turnstones about the rocks near Santa Cruz, in Flores. I was told that in summer they are frequently seen upon the rough pasture-land high up in the mountains. The people say they go there to feed in hot weather; but I suspect they breed there as well, since a lad at Santa Cruz told me that he had shot very young birds. No one, however, that I met with could give me any information about their nesting-habits. The only specimen I procured was a male in full summer-plumage which was shot in June.

39. †Crex pratensis, Bechstein.

Hab. Azores, central group; Europe; Madeira; Bermudas; United States; Greenland.

Mr. J. Dabney showed me a stuffed Corn-Crake, which was killed two or three years previously by flying against a window of his house. I also saw another stuffed specimen in a collection in Angra, said to have been killed in Terceira. The distances to which this short-winged bird occasionally wanders are sufficiently remarkable, since it has occurred in the Bermudas, the United States, and even Greenland.

40. Gallinula chloropus (Linnæus).

Gallinula chloropus, Drouet, Faun. Açor. p. 125.

Hab. Azores, eastern group; N. Africa; Europe; Madeira; Canaries.

41. Fulica atra, Linnæus.

Hab. Azores, eastern group; N. Africa; Europe; Iceland; Madeira.

Both these species are to be found in St. Michael's, on the Lagoa do Fogo. I believe they were originally introduced into the islands, and I am not aware that they are found in any other locality.

42. Anas boschas, Linnæus. "Pato."

Anas boschas, Drouet, Faun. Açor. p. 128.

Hab. Azores, eastern, central, and western groups; N. Africa; Europe; N. America; Madeira; Canaries.

A few Wild Duck are to be found about all the lakes throughout the islands; however, they are very shy. In Flores I saw several in the mountain-lakes and about the marsh, where they breed. In winter they say that several other kinds of Ducks occur; but I only saw those mentioned in this list.

43. Anas crecca, Linnæus. "Mareca."

Anas crecca, Drouet, Faun. Açor. p. 128.

Hab. Azores, eastern, central, and western groups; N. Africa; Europe; Madeira; Canaries; N. America.

Like the last species, a few individuals are found everywhere. The Teal breeds in Flores, but is not quite so common as *A. boschas*.

44. †Œdemia nigra (Linnæus).

Anas nigra, Drouet, Faun. Açor. p. 128.

Hab. Azores, eastern, central, and western groups; Europe.

I saw a black Duck on the lake at the Furnas in St. Michael's, which, I believe, belonged to this species. It is said to occur in all the islands occasionally.

45. †Mergulus alle (Linnæus).

Hab. Central group; N. Europe; N. America.

There is a single specimen in the collection of a gentleman in Terceira, which was killed in the island four or five years ago.

46. Sterna fluviatilis, Naumann. "Carajão."

Sterna hirundo, Morel. Hist. Nat. des Açor. p. 84; Drouet, Faun. Açor. p. 126.

Hab. Azores, eastern, central, and western groups; Europe; Madeira; Canaries; N. America.

This and the following species are the only two real migrants in the Azores. The Common Tern comes about the middle of April, and is to be seen in considerable numbers about the sea-coast and mountain-lakes, departing, I was told, about the middle of September. It breeds on the small islands about the sea-coast.

47. STERNA DOUGALLI, Latham. "Carajão."

Hab. Azores, eastern?, central, and western groups; Africa; Europe; Madeira; Canaries; America.

Mr. S. Dabney, of Fayal, told me that when he was in Flores, about the year 1855, he shot several Terns with pink breasts. During my visit to that island I kept a sharp look-out for them, but did not see any, nor could I find any one amongst the inhabitants who knew the bird, though *S. fluviatilis* was common enough. On my return to Fayal, I one day took a walk to Castello Branco, a large high rock almost detached from the mainland. There were a great many Gulls and Common Terns flying about; and whilst I was watching them as they flew along the side of the cliff, I noticed five or six Roseate Terns amongst them: some of these came within a few yards of me; but I did not shoot at them, as they would have fallen into the sea at the foot of the cliff, where I could not have picked them up. I suspect this species arrives later than the Common Tern, as I afterwards saw several more near the west point of the same island.

48. RISSA TRIDACTYLA (Linnæus).

Larus tridactylus, Morel. Hist. Nat. des Açor. p. 84; Drouet, Faun. Acor. p. 126.

Hab. Azores, eastern, central, and western groups; Madeira; Canaries; N. Europe; N. America.

There were a few Kittiwakes about the harbour of Ponta Delgada when I first arrived; but I did not see them elsewhere. The master of one of the fruit-schooners told me that this and the next species frequently followed their vessels for the whole of the voyage from England. I do not know that it breeds in the Azores.

49. Larus argentatus, Linnæus. "Garça branco."

Larus argentatus, Morel. Hist. Nat. des Açor. p. 84; Drouet, Faun. Açor. p. 126.

Hab. Azores, eastern, central, and western groups; Madeira; Canaries; Europe.

Common everywhere about the sea-coasts and mountain-lakes. Some remain throughout the year, though there are said to be more in summer than in winter. They breed about the coast, and particularly on a small island about a quarter of a mile from the south-west point of Fayal, which in June was quite covered with them.

50. Puffinus major, Faber. "Cargara."

Procellaria puffinus, Morel. Hist. Nat. des Açor. p. 84.

Puffinus cinereus, Drouet, Faun. Açor. p. 127.

Hab. Azorean seas; Europe; America.

To be seen throughout the archipelago. It breeds, in holes in the cliffs, about the end of May. One bird that I shot contained an egg almost ready for exclusion.

51. Puffinus anglorum, Boie. "Stapagado."

Hab. Azorean seas; Europe; America.

Not so numerous as the last species; like it, however, it breeds in holes in the cliffs, in May, and is esteemed by the inhabitants as an article of food. The specimens I procured were wonderfully fat, and the eggs in the ovaries of two females were in an advanced stage.

52. ? Puffinus obscurus (Gmelin). "Frulho."

Hab. Azorean seas; Europe.

People living in the island of Flores told me that there was a smaller bird than the last species, but similar in form, colour, and habits. I hence conclude it is *P. ob-*

scurus. It is said to arrive about the month of March, and to breed in the cliffs. It had reared its young and gone again before I was there, and I did not obtain or even see a specimen; neither did I hear of it in the other islands. The natives frequently bring up young birds of this kind tame, as they afford amusement from their grotesque manner of waddling about.

53. THALASSIDROMA WILSONI, Bonaparte. "Alma de mestre."

Hab. N.E. America; Europe.

On returning from Flores to Fayal we were becalmed for some hours; and as there were a good many Petrels flying about, I took the boat belonging to the schooner and shot some. They were all of this species, nor did I see any other in the archipelago. In flying they carry their legs stretched straight out behind them, and their feet protruded about an inch beyond the tail, producing the effect of two long feathers. I know nothing about this species breeding in the archipelago, though I suspect it does, as it remains throughout the year.

Subjoined is a list* of the birds found in the Azores, with their distribution in Europe and N. Africa, Madeira‡, and the Canaries§, &c.

* The dagger (†) prefixed before the name of a species signifies that it is a straggler, and not a resident.

‡ The Madeiran list is taken from Mr. Vernon Harcourt's paper, "Notice of the Birds of Madeira," in the 'Proceedings of the Zoological Society,' 1861, pp. 141–146; and 'Ann. Nat. Hist.' 2nd ser. vol. xvi. 1855 (V. Harcourt).

§ The Canarian list is taken from the 'Ornithologie Canarienne,' par MM. P. B. Webb, S. Berthelot, et M. Alfred Moquin-Tandon.

	Europe, with N. Africa.	Madeira.	Canaries.	Peculiar to Azores.	
†Tinnunculus alaudarius	*	*	*		
Buteo vulgaris	*	*	*		
Asio otus	*	..	*		
Strix flammea	*	*	*		
Turdus merula	*	*	*		
†Oriolus galbula	*	†*			
Erythacus rubecula, var.	*	*			
Sylvia atricapilla	*	*	*		
Regulus cristatus	*				{ Also Greenland,
Saxicola œnanthe	*	..	†*	..	Labrador, and
Motacilla sulphurea	*	*			Nova Scotia.
†Plectrophanes nivalis	*	..	†*	..	Greenland; North
Fringilla tintillon	..	*	*		America.
Pyrrhula murina	..	..	..	*	
Serinus canarius	..	*	*		
Sturnus vulgaris	*				
Dryobates minor	*				
†Upupa epops	*	*	*		
Columba palumbus	*	*			
—— livia, var.	*	*	*		
Cacabis rufa	*	*			
Coturnix communis	*	*	*		
Ægialites cantianus	*	†*	*		
†Vanellus cristatus	*	†*	†*		
Strepsilas interpres	*	†*	†*		
Ardea cinerea	*	†*	†*		
†—— purpurea	*	†*			
†—— alba	*				
†—— garzetta	*	..	†*		
†—— egretta	*				
†Ardetta minuta	*	†*			
†Botaurus stellaris	*	†*	†*		
†Platalea leucorodia	*	*	†*		
Numenius arquatus	*	†*			
—— phæopus	*	†*	†*		
Scolopax rusticola	*	*	*		
Gallinago media	*	†*	†*		
Tringa maritima	*				
†Crex pratensis	*	†*			
Gallinula chloropus	*	†*	†*		
Fulica atra	*	*			
Anas boschas	*	†*	†*		
—— crecca	*	†*	†*		
†Œdemia nigra	*				
	41	32	26	1	

	European shores.	American shores.	Madeira.	Canaries.	
†Morgulus alle	*	*			
Sterna hirundo	*	*	*	*	
—— dougalli	*	*	†*		
Rissa tridactyla	*	*	†*		
Larus argentatus........	*	*	*		
Puffinus major	*	*	*	*	
—— anglorum	*	*	*	*	
—— obscurus	*	*	*	*	
Thalassidroma wilsoni ..	..	*			
	8	9	7	4	

An analysis of the distribution of the 53 species of birds whose occurrence has been recorded in the Azores, shows us that 15 out of the whole number, or about 28 per cent., consist of accidental stragglers which have not yet become permanently established in the islands. The remaining 38 seem to constitute the resident bird-population.

Of these 38 species, 8 must be regarded as strictly sea-birds; the remaining 30 are either terrestrial or fluviatile species.

91 per cent. of the Azorean birds are also found in Europe, all but three occurring there; but of these, 15 are accidental stragglers, and 8 purely sea-birds. Hence the proportion of non-European to the remainder of the European species is about 10 per cent.

One species, viz. *Pyrrhula murina*, or about 3¼ per cent. after deducting the stragglers and sea-birds, is peculiar.

One other species, viz. *Thalassidroma wilsoni*, remains to be accounted for; but the occurrence of this species, an oceanic wanderer in the North Atlantic, is of no significance.

Very many of the birds of the Azores are also found in the other Atlantic groups; but they have only two species, viz. *Fringilla tintillon* and *Serinus canarius*, in common which are not also European.

The connexion, numerically, with Madeira is closer than with the Canaries.

REPTILIA.

Following M. Drouet I here insert a species of Lizard (*Lacerta dugesi*) which he met with in the island of Graciosa. It did not come under my own observation, as I did not land on that island. The species appears to be peculiar to the Atlantic groups, having hitherto been only found in Madeira and Teneriffe. M. Drouet says it has probably been recently introduced; but still its occurrence in the Azores is significant. Dr. Günther tells me he considers it a good species, and that it is perhaps most nearly allied to the European *L. muralis*. The same high authority says that besides this Lizard a second (*Lacerta gallotti*) is peculiar to the same islands of Madeira and Teneriffe, where, however, there are no other reptiles, nor are there any peculiar freshwater fish.

AMPHIBIA.

Rana esculenta is the only Frog at present known in the Azores; and this, though a recent introduction, has now become fairly established, and is abundant in St. Michael's, and also in some of the central group of islands. It is less common in Flores.

PISCES.

Considering the size of some of the lakes in the Azores, it seems at first rather a remarkable fact that no indigenous freshwater fish is found there; the streams, however, are very small and some of them highly impregnated with sulphur and other mineral substances, which would probably destroy animal life. It is only at the bottom of old craters that any extent of water exists; such are the lakes at the Furnas, the Sete Cidades, and the Lagoa do Fogo. These contain scarcely any aquatic weeds or insects; nevertheless the common Goldfish (*Cyprinus auratus*), introduced some years ago, now swarms in most of them. After a gale of wind some hundreds of dead fish may be seen thrown up by the surf on the shores of these lakes; and their emaciated state shows the difficulty they have in procuring sufficient food.

There is also a species of Eel, two specimens of which were sent me after my return to England—one from Flores and the other from St. Michael's; both were in bad condition. Dr. Günther tells me that he can identify them with *Anguilla fluviatilis* of Europe. M. Drouet says that *Anguilla canariensis* is an inhabitant of the Azores; but he doubtless refers to the former species. I was informed that a Pike is to be found in the Lagoa do Fogo and some other places; but this likewise is a recent introduction, and I did not see it myself.

COLEOPTERA*.

By G. R. Crotch.

The Azores, though not less interesting, have yet received a far less share of attention, as far as their fauna is concerned, than the neighbouring groups of Madeira and the Canaries. The exploration of these, however, is due almost entirely to the laborious and unremitting exertions of Mr. T. V. Wollaston, who has devoted himself to working out the Coleopterous fauna of the Atlantic region with a care and perseverance that, unfortunately, finds too few imitators. The fourth group, viz. the Cape Verde Isles, has also been recently explored by him, and has produced a magnificent series of novelties. It is with considerable pleasure, then, that I am able, through the kindness and liberality of Mr. Godman, to supplement his researches with an enumeration of the Azorean Coleoptera. Our previous knowledge of this group of islands was very limited, but will be found admirably summed up in M. Drouet's 'Eléments de la Faune Açoréenne.' Indeed it is to him and to his companion M. Morelet that we owe any detail of the insects, shells, &c. at all. Of Coleoptera he enumerates fifty-nine, and comments upon their European character, five only being peculiar: these were described in part by M. Tarnier in M. Morelet's 'Notice sur l'histoire naturelle des îles Açores,' and one (*Laparocerus azoricus*) by M. Drouet himself in his 'Coleoptères Açoréens.' The remaining species cited by him are of the most ordinary character, and show certainly the cultivated state of the islands. The material amassed by Mr. Godman was libe-

* This article is a reprint, with alterations, of Mr. Crotch's Paper on Azorean Coleoptera, published in the 'Proceedings of the Zoological Society of London for 1867,' pp. 359-391.

rally placed in my hands to be worked out; and it shows a very great advance upon that of MM. Drouet and Morelet, including as it does 212 species, of which twelve were new to science, and upwards of forty new to the Atlantic district, thus redeeming the fauna from its purely European character. Of the nine islands, three remain practically unvisited—one, indeed (Pico), being probably the best island for characteristic species, being much the most wooded.

Analogy would lead us to put the fauna at, at least, double the present number; and much of the increase would consist of new species, since, in comparing it with that of the other Atlantic groups, it presents some singular features.

Thus, of the 1450 species comprised in the 'Coleoptera Atlantidum,' one-fourth or 25 per cent. are European, one-fourth or 25 per cent. probably geographical races, and one-half or 50 per cent. indigenous. Thus in the combined groups only 350 European species occur, while here we have already 175; hence no great increase of this class can be expected. The proportions here take the form of 83 per cent. European, and about 6½ per cent. indigenous. This is no doubt due to this collection having been made more in cultivated districts and the neighbourhood of towns than under canvas in remote ravines, as Madeira has been worked*; still it shows that the prevailing Atlantic forms are here only scantily represented.

The characteristic genera *Laparocerus, Acalles, Tarphius, Attalus* (all containing eighteen or nineteen species in the other groups) have only solitary representatives. The prevailing genera are *Cryptophagus* (6), *Homalota* (11),

* We spent several weeks at the Furnas in St. Michael's working among the indigenous forests, where, however, our labours were rewarded to a very limited extent.—F. G.

Philonthus (6), *Lithocharis* (5); but they contain almost entirely introduced species. The two new genera of *Rhynchophora, Asynonychus* and *Neocnemis,* barely redeem the general poverty of the fauna; both, however, are very anomalous in their affinities. Two very abundant Madeira forms (*Mesites* and *Dasytes*) are here represented by European species (*M. tardii,* Curt., and *D. nobilis,* Ill.), in place of the cognate species found in the former group; and this is the more singular, as so marked a connexion with Madeira exists in some species*. The conclusions derived from M. Drouet's lists of the other classes accord with some of these deductions: thus the almost total absence of peculiar Vertebrata (no Reptiles) would seem to show that very different conditions from the Canaries must have prevailed. In its land-shells, which afford a good parallel to the insects, out of seventy-six species, one-half are peculiar, one-seventh Atlantic, and one-third European; among these, *Viquesnelia,* peculiar to the Azores and India, though found fossil in the Pyrenees, is the most remarkable.

I now proceed to examine the distribution in detail, separating the species which appear to have been indirectly introduced since the colonization of the islands by man from those which appear to belong more strictly to an indigenous fauna.

Of the 175 European species, 101 are almost certainly introductions, leaving 74 possibly indigenous.

The 101 introduced species may be classified in eight sections as follows:—

* The most striking group in the collection, however, is the *Elateridæ,* with six fine species belonging to as many genera. When we remember that in the Canaries and Madeira this family is represented by the ill-defined and inconspicuous genus *Coptostethus,* Woll., this is very remarkable. Upon examination, however, two appear to be American and two European, thus leaving only two really indigenous.

(1) Cosmopolitan species, which are introduced in articles of commerce, especially provisions, to all parts of the world. These are totally without significance in any fauna, their record depending only on the assiduity with which search is made in warehouses &c. in the sea-ports. Cutting off, therefore, the twelve here enumerated, leaves the real fauna at 200 species.

45. Carpophilus dimidiatus.	108. Anobium paniceum.
46. —— mutilatus.	120. Calandra oryzæ.
57. Silvanus advena.	119. —— granaria.
58. Nausibius dentatus.	159. Tribolium ferrugineum.
72. Corticaria serrata.	160. Tenebrio obscurus.
104. Ptinus testaceus.	161. Alphitobius piceus.

(2) Species also introduced by the medium of commerce, but which may be characterized rather as frequenters of refuse: they are found, for the most part, in the débris of hay- or straw-ricks, about hotbeds, and, indeed, in all vegetable refuse not too rotten †.

41. Sericoderus lateralis.	61. Cryptophagus affinis.
42. Ptenidum apicale.	59. —— cellaris.
48. Nitidula 4-pustulata.	62. —— *punctipennis*.
49. —— colon.	63. —— saginatus.
51. Monotoma 4-foveolata.	64. —— *schmidtii*.
52. —— spinicollis.	66. Atomaria munda.
53. —— quadricollis.	67. Epistemus gyrinoides.
55. Aglenus brunneus.	68. Latridius minutus.
60. Cryptophagus dentatus.	69. —— *nodifer*.

† As an exemplification of the above, it may be interesting to mention that the conditions under which these species thrive seem to have culminated in a small shed used for picking fowls near Horta in Fayal. Under the feathers &c. were found ten of the species here enumerated, which, however, occurred nowhere else in the islands.

71. Corticaria fulva.
74. Typhæa fumata.
75. Mycetæa hirta.
76. Dermestes frischii.
78. Acritus minutus.
79. Carcinops pumilio.
93.*Trox scaber.
153. Blaps similis.
163. Anthicus floralis.
166. Falagria obscura.
186. Philonthus æneus.
187. —— umbratilis.
195. Leptacinus pusillus.
192. Xantholinus punctulatus.
196. Stilicus affinis.
200. Lithocharis ochracea.

(3) Species introduced in old wood &c. in houses.

103. Opilus mollis.
106. Anobium domesticum.
137. Hylotrypes bajulus.
139. Gracilia pygmæa.
138. Clytus 4-punctatus.
141. Leptura fontenayi.

(4) Species inhabiting dung. Here it may be remarked that if islands are dependent on colonization for their Mammalia, their coprophagous insects must also be introduced; special attention should therefore be paid to any new species having these habits. Of course many of the feeders on decaying vegetables will take to dung under certain circumstances.

34. Sphæridium bipustulatum.
36. Cercyon *obsoletum.*
86. *Onthophagus vacca.*
85. —— *taurus.*
87. Aphodius granarius.
88. —— lividus.
167. Aleochara nitida.
168. —— puberula.
177. Homalota atramentaria.
178. —— melanaria.
176. —— nigra.
186. Philonthus sordidus.
188. —— scybalarius.
205. Oxytelus sculptus.
206. —— complanatus.
207. —— nitidulus.

(5) Species introduced with pine trees, as in Madeira.

* See note p. 48.

In the Canaries, where pines are apparently indigenous, the insects are all cognate species.

124. Pissodes notatus.
110. Hylastes *ater*.
111. Hylurgus ligniperda.
211. Homalium pusillum.

(6) Species found in water. Of these there are very few in the islands; why, it is difficult to say; but with the exception of one *Agabus*, which may be new, the rest are the most ordinary European forms. The three new to the Atlantic fauna are probably introduced.

28. Hydroporus *planus*.
29. Colymbetes *pulverosus*.
33. Philhydrus *lividus*.

(7) Species introduced with garden plants &c. about cultivated ground.

98. *Melanotus dichrous*?
99. *Athous obsoletus*?
121. Ceuthorhynchus nigro-terminatus.
128. Hypera variabilis.
126. *Otiorhynchus scabrosus*.
127. —— *sulcatus*.
132. Sitones *flavescens*.
131. —— lineatus.
133. —— gressorius.
134. Bruchus pisi.
143. Psylliodes chrysocephala.
142. *Haltica ampelophaga*.

(8) Of the 86 European species remaining, the following 14 are probably mere recent introductions :—

6. Calathus *mollis*.
22. Bradycellus *distinctus*.
43. Phalacrus coruscus.
44. —— consimilis.
50. Nitidula obsoleta.
77. Anthrenus varius.
80. Saprinus *cærulescens*.
81. Saprinus semistriatus.
109. Ptilinus pectinicornis.
112. Tomicus saxeseni.
113. *Hypoborus ficus*.
145. Coccinella 7-punctata.
148. *Chilocorus bipustulatus*.
150. Scymnus minimus.

The distribution of the whole of the Azorean Coleoptera over the adjoining continent of Europe, together with North Africa and the island groups of the Madeiras and Canaries, may be analyzed as follows :—

The number of European species found in the Azores is 175, or 82·5 per cent. of the whole fauna.

Of these, 97 are also found in both the other Atlantic groups, or 45·7 per cent. of the whole fauna, 55·5 per cent. of the European species.

Of the 97 European species also found in both the other Atlantic groups, 62 are included in the introductions enumerated above, leaving 35 as possibly indigenous, representing, perhaps, the remains of a common fauna. They are as follows :—

2. Blechrus maurus.
4. Pristonychus complanatus.
8. Anchomenus albipes.
9. —— marginatus.
19. Stenolophus teutonus.
24. Tachys 4-signatus.
32. Parnus prolifericornis.
35. Dactylosternum abdominale.
37. Cercyon littorale.
39. —— centromaculatum.
73. Corticaria curta.
82. Saprinus apricarius.
83. —— dimidiatus.
89. Psammodius sabulosus.
91. —— porcicollis.
92. —— cæsus.
105. Mezium sulcatum.
151. Rhizobius litura.
152. Blaps gages.
154. Hegeter tristis.
155. Opatrum hispidum.
165. Anthicus hispidus.
171. Homalota longula.
175. —— coriaria.
180. Habrocercus capillaricornis.
181. Conosomus sericeus.
182. Creophilus maxillosus.
189. Philonthus nigritulus.
193. Xantholinus hesperus.
201. Lithocharis ruficollis.
202. —— debilicornis.
203. Stenus guttula.
204. Platystethus spinosus.
208. Trogophlœus riparius.
209. —— corticinus.

Forty-three species are found in Europe, but not in the other Atlantic islands, or 20 per cent. of the whole fauna, 24½ per cent. of the European species.

Of these 43 species 20 are introductions, leaving 23 indigenous species, viz. :—

3. Licinus *brevicollis.*
5. Calathus *flavipes.*
10. Anchomenus *parumpunctatus.*
12. Pterostichus *vernalis.*
16. Harpalus *ruficornis.*
17. —— *griseus.*
20. Stenolophus *brunnipes.*
21. —— *luridus.*
40. Corytholophus *sublævipennis.*
47. Meligethes *incanus.*
84. Saprinus *rugifrons.*
90. Psammodius *plicicollis.*
102. Dolichosomus *nobilis.*
115. Mesites *tardii.*
135. Bruchus *tristiculus.*
146. Coccinella 11-*punctata.*
147. —— *variabilis.*
172. Homalota *atricilla.*
183. Ocypus *æthiops.*
191. Xantholinus *glabratus.*
198. Lithocharis *ripicola.*
199 —— *apicalis.*
210. Trogophlœus *subtilis.*

Twenty-seven species are common to Europe and Madeira, but are not found in the Canaries, or 13·2 per cent. of the whole fauna, 15·5 per cent. of the European species.

Of these 27 species 15 are introductions, leaving 12 indigenous species, viz. :—

11. Pterostichus nigerrimus.
13. Amara trivialis.
14. Anisodactylus binotatus.
15. Harpalus rotundicollis.
18. —— distinguendus.
25. Bembidium rufescens.
116. Phlœophagus spadix.
136. Bruchus irresectus.
157. Phaleria bimaculata.
170. Homalota luridipennis.
194. Xantholinus linearis.
197. Sunius gracilis.

Eight species are common to Europe and the Canaries, but are not found in Madeira, or 3·75 per cent. of the whole fauna, 4·5 per cent. of the European species.

Of the 8 species common to Europe and the Canaries, but not found in the Madeiras, 4 are introductions, leaving the following 4 indigenous species, viz. :—

31. Gyrinus dejeani.
158. Trachyscelis aphodiodes.
164. Anthicus humilis.
184. Ocypus olens.

We now come to the non-European species, which are 36 in number, or 17 per cent. of the whole fauna.

Of these, 8 are common to the other Atlantic groups, or under 4 per cent. of the whole fauna, 22 per cent. of the non-European species. They are as follows :—

38. Cercyon inquinitum.
56. Læmophlœus clavicollis.
65. Paramecosoma simplex.
70. Corticaria maculosa.
114. Cryphalus aspericollis.
123. Apion chalybeipenne.
144. Psylliodes vehemens.
162. Anaspis proteus.

Eight are found in Madeira, but not in the Canaries, or under 4 per cent. of the whole fauna, 22 per cent. of the non-European species. They are as follows :—

23. Trechichus fimicola.
26. Bembidium schmidtii.
101. Malachius militaris.
117. Phlœophagus tenax.
149. Scymnus durantæ.
169. Homalota obliquepunctata.
190. Philonthus filiformis.
212. Homalium clavicorne.

Three are found in the Canaries, but not in Madeira, or 1·5 per cent. of the whole fauna, 8 per cent. of the non-European species. They are :—

1. Calosoma azoricum.
107. Anobium villosum.
173. Homalota putrescens.

Fourteen are peculiar to the Azores, not having been

found elsewhere, or 6·6 per cent. of the whole fauna, or 39 per cent. of the non-European species, viz. :—

7. Anchomenes aptinoides.
27. Bembidium hesperus.
30. Agabus godmani.
54. Tarphius wollastoni.
94. Heteroderes azoricus.
97. Elastrus dolosus.
100. Attalus miniaticollis.
118. Phlœophagus variabilis.
122. Acalles droueti.
125. Laparocerus azoricus.
129. Asynonychus godmani.
130. Neocnemis occidentalis.
156. Helops azoricus.
179. Xenomma melanocephala.

Three are found in South America, or 1·5 per cent. of the whole fauna, 8 per cent of the non-European species.

One species remains undetermined.

SUMMARY.

European Species.

Common to all the Atlantic Islands . . .	97	
Azores and Madeira only	27	
Azores and Canaries only	8	
Azores only	43	
		—175

Non-European Species.

Common to all the Atlantic Islands . . .	8	
Azores and Madeira only	8	
Azores and Canaries only	3	
Peculiar species	14	
Azores and America	3	
		— 36
Undetermined		1
Total		212

The most remarkable portion of the Azorean fauna is

the presence of those species which it has in common with America. These are:—

95. *Æolus melliculus*.
140. *Tæniotes scalaris*.
96. *Monocrepidius posticus*.

The *Heteroderes azoricus* also is probably a mere modification of an American species, which has succeeded in establishing itself here. The *Tæniotes* also appears thoroughly naturalized. Whether these species owe their introduction to colonization and human intercourse or to natural means must remain an open question. For the former much is to be said. An open and continual communication exists between S. Miguel and Bahia; and Mr. Godman informs me that very large quantities of plants and trees are imported to form gardens. This latter fact may account for the numerous European species also. On the other hand, the occurrence of *Clytus erythrocephalus* on the desolate rocks of the Salvages, where it is not likely to have been thus introduced, suggests that, after all, the Gulf-stream may have been the origin of these peculiar species. This is borne out by the fact that they are all wood-eating species, so that they could readily come in logs in the pupa-state without injury,—and by the fact that the *Heteroderes azoricus* must have been introduced at a period previous to the Portuguese colonization to account for its abundance in several islands and its modified characters. Some light may be thrown on this also by the occurrence of *Cynthia huntera* in the Canaries.

Further, an African connexion is suggested by the very remarkable *Elastrus dolosus*, which has congeners only in Madagascar, but which in external form simulates some Cape *Elaters* so as to be undistinguishable except by a close examination.

A connexion is also found to subsist between Madagascar and the Cape-Verde Islands, two or three species being common to both.

To sum up these affinities numerically, we find that of the 212 species 175 are European, 19 Atlantic, 14 peculiar, 3 American, and 1 undetermined; or that 175 are common to Europe, 140 to Madeira, 116 to the Canaries.

Analysis of the 140 *species found both in the Azores and Madeira.*

Species common to Europe and all the Atlantic Islands	97
Species common to Europe, Azores, and Madeira	27
Species common to Azores, Madeira, and Canaries	8
Species common to Azores and Madeira	8
	140

Analysis of the 116 *species found both in the Azores and Canaries.*

Species common to Europe and all the Atlantic Islands	97
Species common to Europe, Azores, and Canaries	8
Species common to Azores, Madeira, and Canaries	8
Species common to Azores and Canaries	3
	116

The proportions of the families vary a little from those observed in Madeira and the Canaries.

	Azores.	Mad. et Can.
Brachelytra	47	215
Necrophaga	38	219
Rhynchophora	27	282
Geodephaga	27	188
Priocerata	16	135

	Azores.	Mad. et Can.
Cordylocerata . . .	16	64
Heteromera	14	172
Philhydrida	8	29
Pseudotrimera . . .	7	30
Eucerata	5	22
Phytophaga	3	64
Hydradephaga . . .	4	29

The most notable displacements here are the absence of so many of the *Phytophaga,* the lowering of the standard of *Rhynchophora,* always much the largest group in the other islands, and the singular paucity of *Heteromera.* The large development of *Necrophaga* and *Brachelytra* is due to their containing many introduced species. All this seems to show that, on the hypothesis of a connected continent, the fauna of the Azores was drawn from a much more northern source than that of the other islands. This is particularly evinced by the absence of *Heteromera.* The paucity of water-beetles, notwithstanding the great prevalence of rain, is less easily accounted for; but the same occurs in Madeira—where, previously to the destruction of the forests, there must have been water enough, and yet even the universal *Gyrinus dejeani* does not occur there. A more restrained type of fauna is indicated by the solitary representatives of the Atlantic genera (*Tarphius,* &c.), which further south develope numerous forms in each island; it may, indeed, have been that the Azores formed almost the western boundary of land in this direction.

This brief sketch will show how full of interest the subject is, and how much yet remains to be done even in the groups apparently most explored. I shall now enumerate in order the 212 species at present known as inhabitants of these islands.

1. Calosoma azoricum, Heer.

Calosoma azoricum, Woll. Col. Atl. p. 4.

Hab. Azores (*Godman, Drouet*); Lanzarote, Canaries (*Woll.*); Cape-Verde Islands (*Fry*).

Under stones in S. Miguel, Terceira, and Santa Maria, but rarely. Specimens agree precisely with those obtained by Mr. Wollaston from Lanzarote in the Canaries. This species forms the only link between these two groups of islands; it is not, however, confined to them, as it has since been brought from the Cape-Verde Islands. M. Drouet has erroneously identified it with *C. olivieri*, Dej.

2. Blechrus maurus, Sturm.

Blechrus maurus, Woll. Col. Atl. p. 18.

Hab. Azores (*Godman*); Europe; Madeira; Canaries.

Under stones in S. Miguel, not common. None of the examples agree with the allied *B. glabratus*; but this form also will not improbably occur.

3. Licinus brevicollis, Dej.

Hab. Azores (*Godman*); Europe.

Abundant on the sand-hills at Praya in Terceira. New to the Atlantic fauna, but widely spread in the Mediterranean district. Its localization suggests that it may have been imported with ballast.

4. Pristonychus complanatus, Dej.

Pristonychus complanatus, Woll. Col. Atl. p. 27.

Hab. Azores (*Godman*); S. Europe; Madeira; Canaries; St. Helena.

Universal in the various Atlantic islands, including even St. Helena, and common also in parts of South Europe. Specimens are before me from S. Miguel and Flores; but M. Drouet records it from all the islands. This insect also extends to Chili.

5. Calathus flavipes, Payk. (fulvipes, Gyll.).

Hab. Azores (*Drouet*) ; Europe.

Recorded by M. Drouet from all the islands. There is nothing in Mr. Godman's material at all resembling it.

6. C. mollis, Marsh.

Hab. Azores (*Godman, Drouet*) ; Europe.

One specimen from the borders of the Lagoa das Furnas, S. Miguel. M. Drouet records it also from Pico, saying that it is common under stones near the sea. The single specimen before me differs a little from the English form, being larger and with the elytra more deeply striated. The occurrence of two European species only of a genus which almost seems characteristic of the Canaries and Madeira is very remarkable.

7. Anchomenus aptinoides, Tarnier.

Hab. Azores (*Drouet*).

I have not seen this species, described by M. Tarnier from a unique specimen. It would appear to be allied to *A. nichollsii,* Woll., from the Canaries.

8. Anchomenus albipes, Fabr.

Anchomenus albipes, Woll. Col. Atl. p. 35.

Hab. Azores (*Godman*) ; Europe ; Madeira ; Canaries.

Common in damp places in S. Miguel. Also not rare in Madeira ; but in the Canaries it is confined to Fuerteventura.

9. Anchomenus marginatus, Linn.

Anchomenus marginatus, Woll. Col. Atl. p. 35.

Hab. Azores (*Godman, Drouet*) ; Europe ; Madeira ; Canaries.

Margins of the Lagoa das Furnas, S. Miguel, common. M. Drouet adds Terceira. It is common in both the other Atlantic groups.

10. Anchomenus parumpunctatus, Fabr.

Hab. Azores (*Godman, Drouet*); Europe.

A European species new to the Atlantic fauna. It is not rare in S. Miguel, and also (*teste* Drouet) in Fayal and Terceira.

11. Pterostichus nigerrimus, Dej.

Pterostichus nigerrimus, Woll. Col. Atl. p. 40.

Hab. Azores (*Godman*); S.W. Europe; Madeira.

Under stones on the sand-hills at Praya, Terceira, rare. It occurs also in Madeira and South Europe, and is probably a race of *P. aterrimus,* Hb.

12. Pterostichus vernalis, Pz.

Hab. Azores (*Godman, Drouet*); Europe.

New to the Atlantic fauna. M. Drouet records it from all the islands; but I have only seen it from S. Miguel, where it appears to be rare.

13. Amara trivialis, Gyll.

Amara trivialis, Woll. Col. Atl. p. 42.

Hab. Azores (*Godman, Drouet*); Madeira.

S. Miguel and Flores; but also in all the islands, according to M. Drouet. This insect ranges over the whole northern hemisphere.

14. Anisodactylus binotatus, Fabr.

Anisodactylus binotatus, Col. Atl. p. 44.

Hab. Azores (*Godman, Drouet*); Europe; Madeira.

S. Miguel and Terceira. M. Drouet says that it occurs in all the islands.

15. Harpalus (Ophonus) rotundicollis, Fairm.

Ophonus rotundicollis, Woll. Col. Atl. p. 48.

Hab. Azores (*Godman, Drouet*); Europe; Madeira; Salvages.

Common at Angra, Terceira, and Santa Cruz in Flores. M. Drouet records one specimen from S. Miguel. Previously three examples only had been obtained in Madeira, and one from the Salvages; hence its occurrence in the Azores in some numbers is interesting.

16. Harpalus (Pseudophonus) ruficornis, Fabr.

Hab. Azores (*Godman, Drouet*); Europe.

This species abounds in S. Miguel under stones, also in the other islands (*Drouet*). It is new, however, to the Atlantic fauna.

17. Harpalus (Pseudophonus) griseus, Panz.

Hab. Azores (*Godman*); Europe.

Found rarely with the preceding, of which I am disposed to consider it a variety. The only two specimens I have seen are from Terceira and Fayal respectively. They agree with undoubted European specimens; but I cannot think their separation justifiable.

18. Harpalus distinguendus, Dufts.

Harpalus distinguendus, Woll. Col. Atl. p. 46.

Hab. Azores (*Godman*); Europe; Madeira.

This common Madeira insect is probably universally distributed in the Azores. I have seen it from S. Miguel, Terceira, and Fayal.

19. Stenolophus teutonus, Schrank.

Stenolophus teutonus, Woll. Col. Atl. p. 48.

Hab. Azores (*Godman*); Europe; Madeira; Canaries; Mogador.

S. Miguel, Terceira, and Fayal. Probably universal, as in the Canaries.

20. Stenolophus (Acupalpus) brunnipes, Sturm.

Hab. Azores (*Godman*); Europe.

Not uncommon in S. Miguel, Terceira, and Flores; also in Santa Maria, according to M. Drouet. It takes the place of *St. dorsalis,* which is common in Madeira and the Canaries, and of which I regard it as a black variety. It is new to the Atlantic fauna, and its occurrence unmixed with the type form is of considerable interest.

21. Stenolophus (Acupalpus) luridus, Dej.

Hab. Azores (*Godman*); Europe.

On the coast in S. Miguel and Terceira. This is new to the Atlantic fauna; and it is very curious that the pale form should occur, whereas the dark form of the preceding is present.

22. Bradycellus distinctus, Dej.

Hab. Azores (*Godman*); Europe.

One specimen only, from the Lago das Furnas, S. Miguel. It is new to the Atlantic fauna. Compared with English examples, the elytra are more ventricose and have the interstices perceptibly flatter. If further material should show that it is really distinct, I shall propose the name "*azoricus*" for it.

23. Trechichus fimicola, Woll.?

Trechichus fimicola, Woll. Col. Atl. p. 51.

Hab. Azores (*Godman*); Madeira.

One specimen from Fayal. This does not quite agree with Madeiran types in the British Museum, being distinctly paler, and with more faintly striated elytra. Further material can alone decide whether these characters are permanent or not.

24. Tachys 4-signatus, Dufts.

Tachys curvimanus, Woll. Col. Atl. p. 58.

Hab. Azores (*Godman*); S. Europe; Madeira; Canaries.

Not rare in S. Miguel, Terceira, and Fayal. Those from Terceira are paler and more faintly striated. It is common in South Europe, Madeira, and the Canaries; at least I am unable to distinguish between specimens from Spain and others from the latter localities.

25. Bembidium (Ocys) rufescens, Fabr.

Bembidium dubium, Woll. Col. Atl. p. 60.

Hab. Azores (*Godman*); Europe; Madeira.

S. Miguel, Fayal, and Flores; also in Santa Maria (*Drouet*). On carefully comparing it with English specimens and with Mr. Wollaston's type in the British Museum, I am convinced that they should all be referred to one species.

26. Bembidium (Lopha) schmidtii, Woll.

Bembidium schmidtii, Woll. Col. Atl. p. 62.

Hab. Azores (*Godman*); Madeira.

This insect, which assumes a different form in South Europe and the Canaries, here appears to approximate most closely to the Madeiran race; the coloration, however, is darker, the testaceous patches being less developed. It is not common in S. Miguel and Fayal.

27. Bembidium (Leia) hesperus, Crotch.

Bembidium (*Leia*) *hesperus*, P. Z. S. 1867, pp. 369, 385.

Hab. Azores (*Godman*).

Two examples only, under marine rejectamenta at Praya in Terceira. It is most nearly allied to *B. lætum*, Brullé.

28. Hydroporus planus, Fabr.

Hydroporus planus, Woll. Col. Atl. p. 65.

Hab. Azores (*Godman*); Europe; Canaries.

Not rare in ponds in Terceira, Fayal, and Flores. It is

darker than the ordinary English form, but I am unable to detect any tangible differences.

29. Colymbetes (Rhantus) pulverosus, Sturm.

Hab. Azores (*Godman*); Europe.

New to the Atlantic fauna, but is probably introduced; and when one reflects on the introduction of goldfish, it is easy to see that some water-insects at least must have accompanied them. The specimens before me are darker than English ones, a circumstance probably to be accounted for by the method of preservation adopted.

30. Agabus godmanni, Crotch.

Agabus godmanni, P. Z. S. 1867, pp. 370, 385, t. xxiii. f. 3.

Hab. Azores (*Godman*).

This fine species is by no means rare in Terceira, Fayal, and Flores; and it is with some doubt that I have ventured to regard it as new; but it agrees with no published description that I have access to.

31. Gyrinus dejeani, Brullé.

Gyrinus dejeani, Woll. Col. Atl. p. 71.

Hab. Azores (*Godman*); Europe; Canaries.

Common in Flores and Santa Maria, as also in Teneriffe, though not in Madeira. M. Drouet, in his brief list, records no Water-beetles.

32. Parnus prolifericornis, Rossi.

Parnus prolifericornis, Woll. Col. Atl. p. 72.

Hab. Azores (*Godman*); Europe; Madeira; Canaries.

S. Miguel and Santa Maria; also in Graciosa and Flores (*Drouet*).

33. Philhydrus lividus, Forst.

Hab. Azores (*Godman*); Europe.

Not rare in Terceira, but new to the Atlantic fauna, representing the *Ph. melanocephalus* of the other groups.

34. Sphæridium bipustulatum, Fabr.

Sphæridium bipustulatum, Woll. Col. Atl. p. 81.

Hab. Azores (*Godman*); Europe; Madeira.

Common in S. Miguel, Santa Maria, Terceira, and Flores, and is probably, as all the dung species may be presumed to be, universal.

35. Dactylosternum abdominale, Fabr.

Dactylosternum abdominale, Woll. Col. Atl. p. 80.

Hab. Azores (*Godman*); Europe; Madeira; Canaries.

Two specimens, under dung in Fayal. It is somewhat curious that this species should be so rare here, occurring as it does in the Mediterranean district, Madeira, and the Canaries.

36. Cercyon obsoletum, Gyll.

Hab. Azores (*Godman*); Europe.

At Ponta Delgada, S. Miguel, and also in the higher parts of the island, but not common. New to the Atlantic fauna.

37. Cercyon littorale, Gyll.

Cercyon littorale, Woll. Col. Atl. p. 81.

Hab. Azores (*Godman*); Europe; Madeira; Canaries.

Fayal and S. Miguel; one specimen only from each. This species appears to decrease in abundance southwards.

38. Cercyon inquinitum, Woll.

Cercyon inquinitum, Woll. Col. Atl. p. 81.

Hab. Azores (*Godman*); Madeira; Canaries.

One specimen, at Ponte Delgada, S. Miguel. A Madeiran insect, but probably of wider range in reality.

39. Cercyon centromaculatum, Sturm.

Cercyon nigriceps, Woll. Col. Atl. p. 82.

Hab. Azores (*Godman*); Europe; Madeira; Salvages; Canaries.

Flores and Santa Maria, not common. The name "*nigriceps*, Marsh.," has been adopted by some for this species. The description is inapplicable; and Marsham has in his collection placed specimens to represent four different species; hence he could not have had a very clear idea of its characters.

40. Corytholophus sublævipennis, Duv.
Hab. Azores (*Godman*); Europe.
Four or five specimens, from flowers at Horta, Fayal. It seems to agree sufficiently with the European species described by Duval, and is like our common species, but paler and obsoletely punctate.

41. Sericoderus lateralis, Gyll.
Sericoderus lateralis, Woll. Col. Atl. p. 95.
Hab. Azores (*Godman*); Europe; Madeira; Canaries.
S. Miguel and Fayal, in refuse.

42. Ptenidium apicale, Sturm.
Ptenidium apicale, Woll. Col. Atl. p. 101.
Hab. Azores (*Godman*); Europe; Madeira; Canaries.
Fayal, in a shed, among feathers &c., with several other insects of the same imported class.

43. Phalacrus coruscus, Panz.
Phalacrus coruscus, Woll. Col. Atl. p. 103.
Hab. Azores (*Godman*); Europe; Canaries.
One specimen, from Santa Maria. It occurs also in the Canaries, but not in Madeira.

44. Phalacrus (Olibrus) consimilis, Marsh.
Olibrus consimilis, Woll. Col. Atl. p. 105.
Hab. Azores (*Godman*); Europe; Madeira; Canaries.
Abundant in S. Miguel and Fayal; probably an introduced species.

45. Carpophilus dimidiatus, Fabr.
Carpophilus dimidiatus, Woll. Col. Atl. p. 107.
Hab. Azores (*Godman*); Europe; Madeira; Canaries.

From decaying oranges near Ponta Delgada in S. Miguel. A widely ranging species, doubtless introduced.

46. Carpophilus mutilatus, Fabr.
Carpophilus mutilatus, Woll. Col. Atl. p. 107.
Hab. Azores (*Godman*); Europe; Madeira.

Two specimens, with the preceding; both seem nearly cosmopolitan. Introduced with sugar and fruits.

47. Meligethes incanus, Er.
Hab. Azores (*Godman*); Europe.

One, from flowers at Horta in Fayal. It was identified by M. Brisout de Barneville with the above species; and I think the *M. tristis* of Mr. Wollaston's work must also be referred to it.

48. Nitidula 4-pustulata, Fabr.
Nitidula 4-pustulata, Woll. Col. Atl. p. 109.
Hab. Azores (*Godman*); Europe; Madeira.

One specimen, from Ponte Delgada, S. Miguel; clearly introduced.

49. Nitidula (Omosita) colon, L.
Omosita colon, Woll. Col. Atl. p. 110.
Hab. Azores (*Godman*); Europe; Madeira.

In S. Miguel and Fayal, but always in the vicinity of towns.

50. Nitidula (Epuræa) obsoleta, Fabr.
Epuræa obsoleta, Woll. Col. Atl. p. 108.
Hab. Azores (*Godman*); Europe; Madeira.

S. Miguel, Terceira, and Fayal, under bark and refuse &c. In comparison with Madeiran specimens it would seem to be more strongly punctured, and with the thorax just perceptibly more emarginate in front.

51. Monotoma 4-foveolata, Aubé.
Monotoma 4-foveolata, Woll. Col. Atl. p. 119.
Hab. Azores (*Godman*); Europe; Madeira; Canaries.
In the fowl-shed at Horta, Fayal, abundantly.

52. Monotoma spinicollis, Aubé.
Monotoma spinicollis, Woll. Col. Atl. p. 118.
Hab. Azores (*Godman*); Europe; Madeira; Canaries.
Near Horta, Fayal, rarely.

53. Monotoma quadricollis, Aubé.
Monotoma quadricollis, Woll. Col. Atl. p. 119.
Hab. Azores (*Godman*); Europe; Madeira; Canaries.
Ponta Delgada, S. Miguel, one specimen only.

54. Tarphius wollastoni, Crotch.
Hab. Azores (*Godman*).
In dead Euphorbia-stems near Santa Cruz, Flores, not rare. One of the very few remnants of the old laurel-fauna.

55. Aglenus brunneus, Gyll.
Aglenus brunneus, Woll. Col. Atl. p. 129.
Hab. Azores (*Godman*); Europe; Madeira; Canaries.
In the fowl-shed at Horta, Fayal, abundantly.

56. Læmophlœus clavicollis, Woll.
Læmophlœus clavicollis, Woll. Col. Atl. p. 132.
Hab. Azores (*Godman*); Madeira; Canaries.
One specimen, at Ponta Delgada, S. Miguel, but probably more widely distributed.

57. Silvanus advena, Waltl.
Silvanus advena, Woll. Col. Atl. p. 136.

Hab. Azores (*Godman*); Europe; Madeira; Canaries.

At Horta, Fayal, in the fowl-shed, not rare.

58. Nausibius dentatus, Marsh.

Nausibius dentatus, Woll. Col. Atl. p. 134.

Hab. Azores (*Godman*); Europe; Madeira; Canaries.

One specimen, in sugar at Santa Cruz, Flores.

59. Cryptophagus cellaris, Scop.

Cryptophagus cellaris, Woll. Col. Atl. p. 137.

Hab. Azores (*Godman*); Europe; Madeira; Canaries.

In the fowl-shed at Horta, Fayal, rare.

60. Cryptophagus dentatus, Hbst.

Cryptophagus dentatus, Woll. Col. Atl. p. 137.

Hab. Azores (*Godman*); Europe; Madeira; Canaries.

One specimen with the preceding, and one taken by sweeping in S. Miguel.

61. Cryptophagus affinis, Sturm.

Cryptophagus affinis, Woll. Col. Atl. p. 137.

Hab. Azores (*Godman*); Europe; Madeira; Canaries.

Two specimens; one from S. Miguel, the other from Terceira.

62. Cryptophagus punctipennis, Bris.

Hab. Azores (*Godman*); Europe.

One specimen, at Santa Cruz, Flores. This was named for me by M. Brisout himself, and is a species recently described from France.

63. Cryptophagus saginatus, Er.

Cryptophagus saginatus, Woll. Col. Atl. p. 136.

Hab. Azores (*Godman*); Europe; Madeira.

Santa Cruz, Flores, in houses.

64. Cryptophagus schmidtii, Er.?

Hab. Azores (*Godman*); Europe.

One specimen, taken with the preceding, appears to me not to differ from the European species. It is new to the Atlantic fauna.

65. Paramecosoma simplex, Woll.

Paramecosoma simplex, Woll. Col. Atl. p. 140.

Hab. Azores (*Godman*); Madeira; Canaries.

Not rare under refuse in S. Miguel and Fayal. This species has not yet occurred in Europe, though pretty common in all the three groups of islands.

66. Atomaria munda, Er.

Atomaria munda, Woll. Col. Atl. p. 143.

Hab. Azores (*Godman*); Europe; Madeira; Canaries.

In the fowl-shed at Horta, Fayal, abundantly.

67. Epistemus gyrinoides, Marsh.

Epistemus gyrinoides, Woll. Col. Atl. p. 145.

Hab. Azores (*Godman*); Europe; Madeira; Canaries.

With the preceding, also under refuse in S. Miguel.

68. Latridius minutus, L.

Latridius minutus, Woll. Col. Atl. p. 152.

Hab. Azores (*Godman*); Europe; Madeira; Canaries.

One specimen only, at Santa Cruz, Flores. This insect positively swarms in Madeira and the Canaries.

69. Latridius nodifer, Westw.

Hab. Azores (*Godman*).

In decaying oranges at Ponta Delgada, San Miguel, and also at Horta, Fayal, but rarely. This insect has been hitherto confined to England, where it was some years ago of the utmost rarity; now, however, it is universally spread over the country, and in the greatest abundance. It is probably a mere importation into the Azores, but still its presence is not without significance. It has also been found abundantly at Lund in Scandinavia (*Lisle*), and in the Spanish Pyrenees.

70. Corticaria maculosa, Woll.
Corticaria maculosa, Woll. Col. Atl. p. 149.
Hab. Azores (*Godman*); Madeira; Canaries.
At Ponta Delgada, S. Miguel, three specimens only.

71. Corticaria fulva, Com.
Corticaria fulva, Woll. Col. Atl. p. 148.
Hab. Azores (*Godman*); Europe; Madeira; Canaries.
Two specimens only, in S. Miguel and Fayal respectively, and both probably introduced.

72. Corticaria serrata, Gyll.
Corticaria serrata, Woll. Col. Atl. p. 150.
Hab. Azores (*Godman*); Europe; Madeira; Canaries.
One specimen, in the fowl-shed at Horta, Fayal.

73. Corticaria curta, Woll.
Corticaria curta, Woll. Col. Atl. p. 151.
Hab. Azores (*Godman*); Europe; Madeira; Canaries.
S. Miguel and Fayal, under refuse. This species occurs in many parts of Europe, and is scattered in collections as *C. truncatella*, Mannh.

74. Typhæa fumata, L.
Typhæa fumata, Woll. Col. Atl. p. 157.
Hab. Azores (*Godman*); Europe; Madeira; Canaries.
Abundant in S. Miguel and Fayal, under refuse.

75. Mycetæa hirta, Marsh.
Mycetæa hirta, Woll. Col. Atl. p. 156.
Hab. Azores (*Godman*); Europe; Madeira.
Also not rare in S. Miguel and Fayal, in out-houses &c.

76. Dermestes frischii, Kug.

Dermestes frischii, Woll. Col. Atl. p. 160.

Hab. Azores (*Godman*); Europe; Canaries.

S. Miguel, Terceira, and Fayal, in dead fish &c. All the specimens I have examined are referable to this species; but its congener, *D. vulpinus*, must also occur.

77. Anthrenus varius, Fabr.

Anthrenus varius, Woll. Col. Atl. p. 162.

Hab. Azores (*Godman*); Europe; Madeira; Salvages; Canaries.

Very common in flowers in Fayal and Flores, and very variable in size and markings.

78. Acritus minutus, Hbst.

Acritus minutus, Woll. Col. Atl. p. 165.

Hab. Azores (*Godman*); Europe; Madeira; Canaries.

In garden-refuse at Ponta Delgada, S. Miguel, rarely.

79. Carcinops pumilio, Er. (14-striatus, Steph.).

Carcinops 14-striatus, Woll. Col. Atl. p. 172.

Hab. Azores (*Godman*); Europe; Madeira; Canaries.

One specimen, at Horta, Fayal. Mr. Wollaston has employed the Stephensian name to designate this species, which, however, is posterior to Erichson's by five years.

80. Saprinus cærulescens, Ent. H. (semipunctatus, Fab.).

Hab. Azores (*Godman*); Europe.

One specimen, from Ponta Delgada, S. Miguel, has been sent to Mr. Godman since his return. M. Drouet records it from Terceira. It is new to the Atlantic fauna. As the Fabrician insect was different from Herbst's (whose name he quotes), it is impossible to retain the name.

81. Saprinus semistriatus, Scriba (nitidulus, Fabr.).
Saprinus nitidulus, Woll. Col. Atl. p. 171.
Hab. Azores (*Godman*); Europe; Madeira; Canaries.
S. Miguel, Fayal, and Terceira; also common throughout, according to M. Drouet. Scriba's name has eleven years of priority over that of Fabricius.

82. Saprinus apricarius, Er.
Saprinus apricarius, Woll. Col. Atl. p. 168.
Hab. Azores (*Godman*); Europe; Madeira; Canaries.
Abundant in Fayal, under dead fish.

83. Saprinus dimidiatus, Ill.
Hab. Azores (*Drouet, Godman*); Europe.
Abundant with the preceding; M. Drouet also records the species. It must be very close to *S. lobatus*, Woll., if not identical with it.

84. Saprinus rugifrons, Payk.
Hab. Azores (*Drouet*); Europe.
"Under stones on the sea-shore in Terceira."—*Drouet.*
I have not seen any specimens of this species.

85. Onthophagus taurus, Schreb.
Hab. Azores (*Godman*); Europe.
Common in all the islands, and affording a good example of the rapid distribution of an insect in a congenial locality.

86. Onthophagus vacca, Fabr.
Hab. Azores (*Drouet, Godman*); Europe.
One only, from Angra, Terceira. M. Drouet also records one.

87. Aphodius granarius, L.
Aphodius granarius, Woll. Col. Atl. p. 178.
Hab. Azores (*Godman*); Europe; Madeira; Canaries.
S. Miguel, Terceira, and Fayal, abundant.

88. Aphodius lividus, Oliv.

Aphodius lividus, Woll. Col. Atl. p. 178.

Hab. Azores (*Godman*); Europe; Madeira; Canaries.

Not rare in Terceira and Fayal.

89. Psammodius sabulosus, Muls.

Psammodius sabulosus, Woll. Col. Atl. p. 180.

Hab. Azores (*Godman*); Europe; Madeira; Canaries.

One specimen only, from Praya, Terceira; probably, however, not rare.

90. Psammodius plicicollis, Er.

Hab. Azores (*Godman*); Europe.

New to the Atlantic fauna; two specimens were taken at Horta, Fayal.

91. Psammodius porcicollis, Ill.

Psammodius porcicollis, Woll. Col. Atl. p. 180.

Hab. Azores (*Godman*); Europe; Madeira; Canaries.

Abundant near Horta, Fayal.

92. Psammodius cæsus, Panz.

Psammodius cæsus, Woll. Col. Atl. p. 180.

Hab. Azores (*Godman*); Europe; Madeira; Canaries.

Several specimens, from Flores, Terceira, and Fayal.

93. Trox scaber, L.

Trox scaber, Woll. Col. Atl. p. 181.

Hab. Azores (*Godman*); Europe; Madeira.

Abundant in the fowl-shed at Horta, Fayal; but clearly introduced. In Madeira a single specimen only has been noticed.

94. Heteroderes azoricus, Tarn. (atlanticus, Cand.).

Hab. Azores (*Godman*).

Very abundant under stones, and probably universal. I have seen specimens from S. Miguel, Flores, Terceira,

Fayal, and Corvo. M. Drouet also records it from Santa Maria. Mr. E. W. Janson, to whom I am indebted for a careful examination of this and the following *Elateridæ*, informs me that it is nearly allied to *H. rufangulus*, Gyll., of Brazil. This and the remaining *Elateridæ* are entirely new to the Atlantic fauna.

95. ÆOLUS MELLICULUS, Cand. (MORELETI, Tarn.).

Hab. Azores (*Godman*); S. America.

Mr. Janson informs me that the specimens taken by Mr. Godman are not distinguishable from the original Dejeanian types of the above South-American species. It is very widely spread from Carthagena to Buenos Ayres, according to Candèze. The *Æ. moreleti*, Tarnier (1860), is slightly different in coloration; but of the few specimens before me no two are precisely alike. They were all taken at Horta, Fayal, under dead weeds.

96. MONOCREPIDIUS POSTICUS, Erichs.

Hab. Azores (*Godman*); S. America.

A single specimen taken by Mr. Godman in Fayal is referable, as Mr. Janson informs me, to the above common Brazilian species.

97. ELASTRUS DOLOSUS, Crotch.

Hab. Azores (*Godman*).

One specimen only, in S. Miguel. This is probably the *Ampedus*, sp. ?, of M. Drouet's catalogue.

98. MELANOTUS DICHROUS, Erichs. ?

Hab. Azores (*Godman*); S. Europe.

Mr. Janson refers a single specimen taken in Santa Maria to this South-European species.

99. ATHOUS OBSOLETUS, Ill. ?

Hab. Azores (*Godman*); S. Europe.

Three specimens taken in a garden at Ponta Delgada, S. Miguel, appear to be identical with the above species, which, though an inhabitant of South Europe, is new to the Atlantic fauna.

100. Attalus miniatocollis, Tarnier.

Hab. Azores (*Godman*).

Terceira and Fayal, common on flowers. M. Drouet records it from Santa Maria. It is very closely allied to the Canarian *A. ruficollis*, Woll.

101. Malachius militaris, Woll.

Malachius militaris, Woll. Col. Atl. p. 195.

Hab. Azores (*Godman*); Madeira.

A single female specimen, from flowers near the Furnas. It differs from Madeiran examples, communicated to me by Mr. Wollaston, by the form of the thorax, which in them is slightly narrowed behind and sinuated, whereas in this it is nearly quadrate; but the punctuation is nearly identical.

102. Dolichosomus nobilis, Ill.

Hab. Azores (*Godman*); Europe.

Probably universal, as Mr. Godman brought it from Terceira, S. Miguel, Fayal, Flores, and Corvo M. Drouet also records it from Santa Maria. Its occurrence is somewhat remarkable, since in Madeira it is represented by the nearly allied *D. illustris*, Woll.

103. Opilus mollis, Linn.

Opilus mollis, Woll. Col. Atl. p. 208.

Hab. Azores (*Godman*); Europe; Madeira.

One specimen, from a house in Ponta Delgada, S. Miguel. M. Drouet speaks of it as common.

104. Ptinus testaceus, Oliv.

Ptinus testaceus, Woll. Col. Atl. p. 213.

Hab. Azores (*Godman*); Europe; Madeira; Canaries.

Two specimens in the fowl-shed at Horta, Fayal.

105. Mezium sulcatum, Fabr.

Mezium sulcatum, Woll. Col. Atl. p. 214.

Hab. Azores (*Godman*); Europe; Madeira; Canaries.

In Terceira, Fayal, and Santa Maria, not rare. This species is probably universal, being very abundant in Madeira and the Canaries, where it seems truly indigenous.

106. Anobium domesticum, Fourc. (striatum, Oliv.).

Anobium striatum, Woll. Col. Atl. p. 227.

Hab. Azores (*Godman*); Europe; Madeira; Canaries.

Terceira and Santa Maria, in houses. M. Drouet says that it occurs throughout the group.

107. Anobium villosum, Brullé?

Anobium villosum, Woll. Col. Atl. p. 225.

Hab. Azores (*Godman*); Canaries.

M. Drouet records *A. tomentosum* as common throughout the group, referring probably to the above Canarian species; but I have seen no specimens of it as yet.

108. Anobium paniceum, Linn.

Anobium paniceum, Woll. Col. Atl. p. 227.

Hab. Azores (*Godman*); Europe; Madeira; Canaries.

From Flores only; but doubtless universal in towns.

109. Ptilinus pectinicornis, Linn.

Ptilinus pectinicornis, Woll. Col. Atl. p. 229.

Hab. Azores (*Godman*); Europe; Madeira.

In houses at the Furnas and other places in S. Miguel, but clearly introduced.

110. Hylastes ater, Fabr.

Hab. Azores (*Godman*); Europe.

From pine trees at Horta, Fayal. New to the Atlantic fauna.

111. HYLURGUS LIGNIPERDA, Fabr.

Hylurgus ligniperda, Woll. Col. Atl. 250.

Hab. Azores (*Godman*); Europe; Madeira; Canaries.

With the preceding, but more abundant.

112. TOMICUS SAXESENI, Ratz.

Tomicus saxeseni, Woll. Col. Atl. p. 237.

Hab. Azores (*Godman*); Europe; Madeira; Canaries.

In abundance in one tree in S. Miguel.

113. HYPOBORUS FICUS, Er.

Hypoborus ficus, Woll. Col. Atl. p. 248.

Hab. Azores (*Godman*); S. Europe; Madeira.

Abundant in a dead fig-tree at Horta, Fayal. This is evidently introduced from the Mediterranean, where it takes the place of the Canarian genus *Liparthrum.*

114. CRYPHALUS ASPERICOLLIS, Woll.

Cryphalus aspericollis, Woll. Col. Atl. p. 239.

Hab. Azores (*Godman*); Madeira; Canaries.

With the preceding, but more rarely. This pretty little insect is universal in the Atlantic groups, extending even to St. Helena.

115. MESITES TARDII, Curt.

Hab. Azores (*Godman*); Europe.

From *Erica*-stems in S. Miguel, and afterwards from a dead *Euphorbia* in Flores; the latter locality, however, must be merely accidental. After a very careful comparison with English and Irish specimens, I am unable to detect any difference between them, improbable as such identity would at first appear to be.

116. Phlœophagus spadix, Hbst.

Phlœophagus sulcipennis, Woll. Col. Atl. p. 253.

Hab. Azores (*Godman*); Europe; Madeira.

Under rubbish at Horta, Fayal; it occurs also in Madeira, and was described by Mr. Wollaston, who, however, expressed his opinion that it might prove to be only a geographical state of *P. spadix*. After comparing a large number of specimens, I think the characters pointed out by him shade away insensibly.

117. Phlœophagus tenax, Woll.

Phlœophagus tenax, Woll. Col. Atl. p. 253.

Hab. Azores (*Godman*); Madeira.

Taken pretty abundantly in an *Erica*-stem at the Furnas, S. Miguel, also in Fayal. The specimens before me differ from Madeiran types sent me by Mr. Wollaston in being less evidently punctate and more rugose on the elytra. The sculpture of this genus, however, is liable to considerable variation in this respect.

118. Phlœophagus variabilis, Crotch.

Hab. Azores (*Godman*).

Common in S. Miguel, Fayal, Flores, and Corvo, and assuming a slightly different form in each island. It feeds on fig-trees and Euphorbias, in a manner analogous to the *P. laurineus*, Woll., and, like that, is more sparingly punctured when found on Euphorbias.

119. Calandra granaria, L.

Sitophilus granarius, Woll. Col. Atl. p. 264.

Hab. Azores (*Godman*); Europe; Madeira; Canaries.

Two specimens from Terceira.

120. Calandra oryzæ, L.

Sitophilus oryzæ, Woll. Col. Atl. p. 265.

Hab. Azores (*Godman*); Europe; Madeira; Canaries.

Very abundant in grain in S. Miguel, Terceira, and Fayal.

121. Ceuthorhynchus nigroterminatus, Woll.
Ceuthorhynchus nigroterminatus, Woll. Col. Atl. p. 268.
Hab. Azores (*Godman*); Europe; Madeira; Canaries.
Not rare on flowers in S. Miguel and Fayal. It occurs also in Europe, and even in England.

122. Acalles droueti, Crotch.
Hab. Azores (*Godman*).
This beautiful species was taken in tolerable numbers from some decayed Euphorbia-stems in Flores.

123. Apion chalybeipenne, Woll.
Apion chalybeipenne, Woll. Col. Atl. p. 292.
Hab. Azores (*Godman*); Madeira; Canaries.
By sweeping in Fayal and Flores, not rare.

124. Pissodes notatus, Fabr.
Pissodes notatus, Woll. Col. Atl. p. 298.
Hab. Azores (*Godman*); Europe; Madeira.
From pine trees at Horta, Fayal; but evidently introduced.

125. Laparocerus azoricus, Drouet.
Hab. Azores (*Drouet, Godman*).
M. Drouet described this from specimens from Fayal; Mr. Godman, however, found it abundantly in S. Miguel under stones. It represents a curious form of the genus, differing from the Canarian species considerably in aspect.

126. Otiorhynchus scabrosus, Marsh.
Hab. Azores (*Godman*); Europe.
Beaten from hedges at Ponta Delgada, S. Miguel. New to the Atlantic fauna.

127. Otiorhynchus sulcatus, Fabr.

Hab. Azores (*Godman, Drouet*); Europe.

One specimen, from the Lagoa das Furnas. M. Drouet records it from Terceira. It is new to the Atlantic fauna, though doubtless introduced.

128. Hypera variabilis, Hb.

Hypera murina, Woll. Col. Atl. p. 305.

Hab. Azores (*Godman*); Europe; Madeira; Canaries.

Under refuse in Terceira, not common.

129. Asynonychus godmanni, Crotch.

Asynonychus godmanni, Crotch, P.Z.S. 1867, pp. 378, 389, t. xxiii. f. 7.

Hab. Azores (*Godman*).

Two specimens of this new and interesting form were beaten from brambles at Horta, Fayal.

130. Neocnemis occidentalis, Crotch.

Neocnemis occidentalis, Crotch, P.Z.S. 1867, pp. 378, 389, t. xxiii. f. 7.

Hab. Azores (*Godman*).

One specimen only was swept from flowers in Santa Maria by Mr. Brewer.

131. Sitona lineatus, L.

Sitona lineatus, Woll. Col. Atl. p. 336.

Hab. Azores (*Godman, Drouet*); Europe; Madeira; Canaries.

In S. Miguel, Terceira, and Fayal, common. M. Drouet records it also from Pico.

132. Sitona flavescens, Marsh.

Hab. Azores (*Godman*); Europe.

One specimen only, by sweeping in Santa Maria. This is a curious variety with a triangular pale sutural patch, not rare in South Europe. It is, however, new to the Atlantic fauna.

133. Sitona gressorius, Fabr.
Sitona gressorius, Woll. Col. Atl. p. 334.
Hab. Azores (*Godman*); Europe; Madeira; Canaries.
Abundant in Terceira and Fayal, in cultivated grounds.

134. Bruchus pisi, L.
Bruchus pisi, Woll. Col. Atl. p. 340.
Hab. Azores (*Godman*); Europe; Madeira; Canaries.
Abundant in gardens in S. Miguel and Flores.

135. Bruchus tristiculus, Schön.
Bruchus azoricus, Crotch, P. Z. S. 1867, pp. 378, 390.
Hab. Azores (*Godman*); Algeria.
In S. Miguel, Terceira, and Fayal, not rare in flowers. It is also found in Algeria.

136. Bruchus irresectus, Schön.
Bruchus breweri, Crotch, P. Z. S. 1867, pp. 379, 389.
Bruchus subellipticus, Woll. Col. Atl. p. 341.
Hab. Azores (*Godman*); Algeria; Madeira.
Two specimens were taken by Mr. Brewer in Santa Maria. It occurs also in Algeria and Madeira, whence it has been described by Mr. Wollaston as *B. subellipticus.*

137. Hylotrypes bajulus, L.
Hylotrypes bajulus, Woll. Col. Atl. p. 343.
Hab. Azores (*Godman*); Europe; Madeira; Canaries.
Several specimens, in and about houses at Ponta Delgada, S. Miguel.

138. Clytus 4-punctatus, Fabr.
Clytus webbii, Woll. Col. Atl. p. 346.
Hab. Azores (*Godman*); Europe; Madeira.
In a chestnut stump at Ponta Delgada, S. Miguel. It was accompanied by the variety *C. griseus,* where the ochreous pubescence is replaced by pale grey. The occurrence of this species here throws light on the question

discussed by Mr. Wollaston as to its occurrence in the Canaries. It would appear not improbable that Mr. Webb did in reality obtain specimens either in Madeira or in the Canaries; but the *C. webbii* is obviously a mere variety of the type form.

139. Gracilia pygmæa, Fabr.

Gracilia pygmæa, Woll. Col. Atl. p. 348.

Hab. Azores (*Godman*); Europe; Madeira; Canaries.

One specimen, in a house at Horta, Fayal.

140. Tæniotes scalaris, Fabr.

Hab. Azores (*Godman, Drouet*); Brazil.

This fine Brazilian species appears to have made good its position in these islands, where it does considerable damage to the fig-trees. It is most abundant in S. Miguel; but M. Drouet states that it occurs also in Fayal and Terceira.

141. Leptura fontenayi, Muls.

Leptura, sp. ?, Crotch, P. Z. S. 1867, p. 379.

Hab. Azores (*Godman*); Europe.

Found also in Southern Europe, but new to the Atlantic fauna.

142. Haltica ampelophaga, Guér.

Hab. Azores (*Godman*); Europe.

Abundant on the vines in Santa Maria, but has been clearly introduced; it is, however, curious that it should not have found its way to Madeira or the Canaries.

143. Psylliodes chrysocephala, L.

Psylliodes chrysocephala, Woll. Col. Atl. p. 372.

Hab. Azores (*Godman*); Europe; Madeira.

Apparently nearly universal, being found in S. Miguel, Terceira, Fayal, and Flores.

144. Psylliodes vehemens, Woll.

Psylliodes vehemens, Woll. Col. Atl. p. 373.

Hab. Azores (*Godman*); Madeira; Canaries.

Not rare in Fayal, and probably in the other islands also. It is very abundant in the other Atlantic groups.

145. Coccinella 7-punctata, L.

Coccinella 7-punctata, Woll. Col. Atl. p. 378.

Hab. Azores (*Godman*); Europe; Madeira; Canaries.

One specimen only, from Santa Maria.

146. Coccinella 11-punctata, L.

Hab. Azores (*Godman, Drouet*); Europe.

From S. Miguel, Terceira, and Flores; also in Santa Maria, according to Drouet.

147. Coccinella variabilis, Fabr.

Hab. Azores (*Drouet*); Europe.

M. Drouet states this insect to be common throughout the group; no trace of it, however, exists in the material now before me; yet it is impossible to believe that so well-known a species can have been confused with any thing else.

148. Chilocorus bipustulatus, L.

Hab. Azores (*Godman*); Europe.

One specimen only, from Santa Maria.

149. Scymnus durantæ, Woll.

Scymnus durantæ, Woll. Col. Atl. p. 380.

Hab. Azores (*Godman*); Madeira.

Not rare on flowers in Terceira and Fayal. This species has been hitherto considered peculiar to Madeira, and is represented by a cognate form in the Canaries.

150. Scymnus minimus, Rossi.

Scymnus minimus, Woll. Col. Atl. p. 382.

Hab. Azores (*Godman*); Europe; Madeira; Canaries.

A few specimens from Fayal and Santa Maria.

151. Rhizobius litura, Fabr.

Rhizobius litura, Woll. Col. Atl. p. 383.

Hab. Azores (*Godman*); Europe; Madeira; Canaries.

Very abundant in S. Miguel, Fayal, and Terceira, and presenting, as usual, considerable variation in colour.

152. Blaps gages, L.

Blaps gages, Woll. Col. Atl. p. 401.

Hab. Azores (*Godman*); Europe; Madeira; Canaries.

S. Miguel and Fayal, in gardens, cellars, &c., not rare. This species has also been found on the Salvages.

153. Blaps similis, Latr.

Blaps similis, Woll. Col. Atl. p. 402.

Hab. Azores (*Godman*); Europe; Madeira; Canaries.

Very common in S. Miguel, Fayal, and Flores.

154. Hegeter tristis, Fabr.

Hegeter tristis, Woll. Col. Atl. p. 395.

Hab. Azores (*Godman*); N. and W. Africa; Madeira; Canaries.

S. Miguel, Terceira, and Fayal, but rare. This insect is excessively abundant in Madeira and the Canaries.

155. Opatrum hispidum, Brullé.

Opatrum hispidum, Woll. Col. Atl. p. 413.

Hab. Azores (*Godman*); S. Europe; Madeira; Canaries.

Abundant in S. Miguel, Terceira, and Fayal, and probably universal.

156. Helops azoricus, Crotch.

Helops azoricus, Crotch, P. Z. S. 1867, pp. 380, 390.

Hab. Azores (*Godman*).

Under the bark of a poplar tree at the Furnas, S. Miguel. This is very near one of the Madeiran species, but not, I think, identical with it.

157. Phaleria bimaculata, Herbst.

Phaleria bimaculata, Woll. Col. Atl. p. 417.

Hab. Azores (*Godman*); S. Europe; Salvages.

Abundant under dead fish at Horta, Fayal. M. Drouet records the *P. cadaverina* from Terceira and S. Miguel, but he evidently means the present species.

158. Trachyscelis aphodioides, Latr.

Trachyscelis aphodioides, Woll. Col. Atl. p. 416.

Hab. Azores (*Godman*); Europe; Canaries.

One specimen only, on the sea-shore at Horta, Fayal.

159. Tribolium ferrugineum, Fabr.

Tribolium ferrugineum, Woll. Col. Atl. p. 420.

Hab. Azores (*Godman, Drouet*); Europe; Madeira; Canaries.

Recorded by M. Drouet from Santa Maria; it is a cosmopolitan insect, and hence of little interest.

160. Tenebrio obscurus, Fabr.

Tenebrio obscurus, Woll. Col. Atl. p. 424.

Hab. Azores (*Godman*); Europe; Madeira; Canaries.

In S. Miguel, Santa Maria, and Graciosa, in bake-houses &c.

161. Alphitobius piceus, Oliv.

Alphitobius piceus, Woll. Col. Atl. p. 419.

Hab. Azores (*Godman*); Europe; Madeira; Canaries.

One specimen, with the preceding, from S. Miguel.

162. Anaspis proteus, Woll.

Anaspis proteus, Woll. Col. Atl. p. 440.

Hab. Azores (*Godman*); Madeira; Canaries.

Abundant on flowers at Fayal. M. Drouet records *A. humeralis*, Fabr., from Santa Maria and S. Miguel; but I feel no doubt that he alludes to this species, which swarms in Madeira and the Canaries, and is very closely allied to a South-European form, if not identical with it.

163. Anthicus floralis, L.

Anthicus floralis, Woll. Col. Atl. p. 443.

Hab. Azores (*Godman*); Europe; Madeira; Canaries.

One specimen of this common insect has occurred in Fayal.

164. Anthicus humilis, Laf.

Anthicus humilis, Woll. Col. Atl. p. 444.

Hab. Azores (*Godman*); Europe; Canaries.

Not rare round the lake at Praya, Terceira.

165. Anthicus hispidus, Rossi.

Anthicus hispidus, Woll. Col. Atl. p. 444.

Hab. Azores (*Godman*); Europe; Madeira; Canaries.

Under refuse in S. Miguel and Flores.

166. Falagria obscura, Grav.

Falagria obscura, Woll. Col. Atl. p. 452.

Hab. Azores (*Godman*); Europe; Madeira; Canaries.

Not common, but found in S. Miguel, Fayal, and Santa Maria.

167. Aleochara nitida, Grav.

Aleochara nitida, Woll. Col. Atl. p. 475.

Hab. Azores (*Godman*); Europe; Madeira; Canaries.

Abundant in dung in S. Miguel, Terceira, Fayal, and Flores.

168. Aleochara puberula, Klug.

Aleochara puberula, Woll. Col. Atl. p. 473.

Hab. Azores (*Godman*); S. Europe; Madeira; Canaries.

Only one specimen, from dung in Fayal.

169. Homalota obliquepunctata, Woll.

Homalota obliquepunctata, Woll. Col. Atl. p. 461.

Hab. Azores (*Godman*); Madeira.

Several specimens, from the margins of the Lagoa das

Farnas. It agrees almost exactly with the Madeiran specimens; but the oblique markings are less evident.

170. Homalota luridipennis, Mannh.
Homalota luridipennis, Woll. Col. Atl. p. 462.
Hab. Azores (*Godman*); Europe; Madeira.
One specimen, taken near the Furnas in S. Miguel.

171. Homalota longula, Heer.
Homalota longula, Woll. Col. Atl. p. 464.
Hab. Azores (*Godman*); Europe; Madeira; Canaries.
Found, but very rarely, in the bed of a stream in Fayal.

172. Homalota atricilla, Er. (flavipes, Thoms.).
Hab. Azores (*Godman*); Europe.
One specimen from the coast at Ponta Delgada, S. Miguel. This species is new to the Atlantic fauna, and is interesting as showing the wide distribution of these seaweed-infesting forms.

173. Homalota putrescens, Woll.?
Homalota putrescens, Woll. Col. Atl. p. 470.
Hab. Azores (*Godman*); Canaries.
From Flores, under refuse. These are not in good condition, but appear to be near Mr. Wollaston's species. They will almost certainly prove to be European also.

174. Homalota ——?
S. Miguel, under refuse.

175. Homalota coriaria, Kraatz?
Homalota coriaria, Woll. Col. Atl. p. 469.
Hab. Azores (*Godman*); Europe; Madeira; Canaries.
Three specimens, from S. Miguel. All these specimens, as well as those of the preceding species, are females, and I am unable to identify them satisfactorily. Both insects, however, appear to belong to European forms.

176. Homalota nigra, Kraatz?
Homalota nigra, Woll. Coll. Atl. p. 466.
Hab. Azores (*Godman*); Europe; Canaries.

This little species, which is not rare in dung, appears to agree with the Canarian specimens referred to *H. nigra,* Kr., by Mr. Wollaston.

177. Homalota atramentaria, Gyll.
Homalota atramentaria, Woll. Col. Atl. p. 467.
Hab. Azores (*Godman*); Europe; Madeira; Canaries.

Not rare in dung in S. Miguel, Fayal, and Flores, and probably universal.

178. Homalota melanaria, Sahlb.
Homalota melanaria, Woll. Col. Atl. p. 471.
Hab. Azores (*Godman*); Europe; Madeira; Canaries.

Abundant in dung in Terceira, Fayal, and S. Miguel.

179. Xenomma melanocephala, Crotch.
Xenomma melanocephala, Crotch, P.Z.S. 1867, pp. 382, 390.
Hab. Azores (*Godman*).

Two specimens from rubbish in S. Miguel. It is allied to the other Atlantic species, but is abundantly distinct from them.

180. Habrocerus capillaricornis, Grav.
Habrocerus capillaricornis, Woll. Col. Atl. p. 481.
Hab. Azores (*Godman*); Europe; Madeira; Canaries.

Two specimens, from vegetable refuse in S. Miguel. This would appear to be a remnant of the old laurel-fauna.

181. Conosoma sericeum, Latr. (pubescens, Payk.).
Conosoma pubescens, Woll. Col. Atl. p. 478.
Hab. Azores (*Godman*); Europe; Madeira; Canaries.

A single mutilated specimen, from a Euphorbia-stem in Flores.

182. Creophilus maxillosus, L.

Creophilus maxillosus, Woll. Col. Atl. p. 487.

Hab. Azores (*Godman, Drouet*); Europe; Madeira; Canaries.

Local, but abundant in some places in S. Miguel and Fayal. M. Drouet records it also from Flores and Graciosa.

183. Ocypus æthiops, Waltl.

Staphylinus hesperus, Crotch, P. Z. S. 1867, pp. 383, 391.

Hab. Azores (*Godman*); Europe.

Abundant under stones near Terceira. It appears to be allied to a Cape species.

184. Ocypus olens, Müll.

Ocypus olens, Woll. Col. Atl. p. 487.

Hab. Azores (*Godman*); Europe; Canaries.

Common throughout the group, as it is also in the Canaries. Its absence from Madeira is a very curious and important fact.

185. Philonthus æneus, Rossi.

Philonthus æneus, Woll. Col. Atl. p. 490.

Hab. Azores (*Godman*); Europe; Madeira; Salvages.

Rare in S. Miguel and Fayal, and probably a mere introduction.

186. Philonthus sordidus, Grav.

Philonthus sordidus, Woll. Col. Atl. p. 491.

Hab. Azores (*Godman*); Europe; Madeira; Canaries.

In vegetable refuse and dung in S. Miguel and Fayal.

187. Philonthus umbratilis, Grav.

Philonthus umbratilis, Woll. Col. Atl. p. 490.

Hab. Azores (*Godman*); Europe; Madeira; Canaries.

Tolerably common in S. Miguel and Fayal. M. Drouet records *P. ventralis*; but I am inclined to imagine he had the present species in view, notwithstanding the discrepancy in the thoracic punctures.

188. Philonthus scybalarius, Nordm.

Philonthus scybalarius, Woll. Col. Atl. p. 492.

Hab. Azores (*Godman*); Europe; Madeira; Canaries; Ascension.

Two specimens taken in Fayal are referable to this species.

189. Philonthus nigritulus, Grav.

Philonthus nigritulus, Woll. Col. Atl. p. 494.

Hab. Azores (*Godman*); Europe; Madeira; Canaries.

Very abundant in damp places in S. Miguel, Terceira, Flores, and Fayal.

190. Philonthus filiformis, Woll.

Philonthus proximus, Crotch, P. Z. S. 1867, p. 383.

—— *filiformis*, Woll. Col. Atl. p. 496.

Hab. Azores (*Godman*); Madeira; Canaries.

A single specimen, from a mountain-stream in Fayal.

191. Xantholinus glabratus, Grav.

Hab. Azores (*Drouet*); Europe.

According to M. Drouet this was found by M. Hartung under stones in Graciosa. If this indication be correct, it is a species new to the Atlantic fauna. Possibly he may allude to the analogously coloured *X. marginalis*, Woll., hitherto found only in the Canaries.

192. Xantholinus punctulatus, Payk.

Xantholinus punctulatus, Woll. Col. Atl. p. 497.

Hab. Azores (*Godman*); Europe; Madeira; Canaries.

Not rare in S. Miguel and Terceira.

193. Xantholinus hesperius, Er.

Xantholinus hesperius, Woll. Col. Atl. p. 497.

Hab. Azores (*Godman*); Europe; Madeira; Canaries.

In San Miguel and Fayal, under refuse.

194. Xantholinus linearis, Oliv.
Xantholinus linearis, Woll. Col. Atl. p. 497.
Hab. Azores (*Godman*); Europe; Madeira.
With the preceding, and also in Terceira.

195. Leptacinus pusillus, Steph. (linearis, Grav.).
Leptacinus linearis, Woll. Col. Atl. p. 498.
Hab. Azores (*Godman*); Europe; Madeira; Canaries.
Under refuse at Ponta Delgada, S. Miguel. Gravenhorst's name is inapplicable, having been adopted erroneously from Olivier.

196. Stilicus affinis, Er.
Stilicus affinis, Woll. Col. Atl. p. 503.
Hab. Azores (*Godman*); Europe; Madeira; Canaries.
With the preceding, but rarer.

197. Sunius gracilis, Payk. (angustatus, Pk.).
Sunius angustatus, Woll. Col. Atl. p. 509.
Hab. Azores (*Godman*); Europe; Madeira.
Under stones in S. Miguel, Fayal, and Flores, not rare. Paykull's name "*angustatus*" was preoccupied; hence we should use the name he subsequently proposed for it.

198. Lithocharis ripicola, Kraatz.
Hab. Azores (*Godman*); Europe.
New to the Atlantic fauna. One specimen only, from S. Miguel.

199. Lithocharis apicalis, Kraatz.
Hab. Azores (*Godman*); Europe.
Two specimens, at Horta, Fayal, under refuse. Also new to the fauna.

200. Lithocharis ochracea, Grav.
Lithocharis ochracea, Woll. Col. Atl. p. 506.

Hab. Azores (*Godman*); Europe; Madeira; Canaries.

One example only, from Fayal, with the preceding.

201. LITHOCHARIS RUFICOLLIS, Kraatz.

Lithocharis tricolor, Woll. Col. Atl. p. 507.

Hab. Azores (*Godman*); Europe; Madeira.

Not rare in S. Miguel, under stones &c. Mr. Wollaston has used Marsham's name for this species; but his description is quite valueless, and the name, moreover, adopted erroneously from Fabricius.

202. LITHOCHARIS DEBILICORNIS, Woll.

Lithocharis debilicornis, Woll. Col. Atl. p. 508.

Hab. Azores (*Godman*); Europe; Madeira; Canaries.

Not rare near Ponta Delgada, under refuse.

203. STENUS GUTTULA, Müll.

Stenus guttula, Woll. Col. Atl. p. 511.

Hab. Azores (*Godman*); Europe; Madeira; Canaries.

At the roots of grass round the mountain-streams in S. Miguel, but rare.

204. PLATYSTETHUS SPINOSUS, Er.

Platystethus spinosus, Woll. Col. Atl. p. 515.

Hab. Azores (*Godman*); Europe; Madeira; Canaries.

Under marine rejectamenta in Terceira and Fayal. These agree with the depauperated phase found in Madeira.

205. OXYTELUS SCULPTUS, Grav.

Oxytelus sculptus, Woll. Col. Atl. p. 516.

Hab. Azores (*Godman*); Europe; Madeira; Canaries.

Very common in dung in S. Miguel, Terceira, Fayal, and Flores.

206. OXYTELUS COMPLANATUS, Er.

Oxytelus complanatus, Woll. Col. Atl. p. 517.

Hab. Azores (*Godman*); Europe; Madeira; Canaries.

With the preceding, and even more abundant.

207. Oxytelus nitidulus, Grav.

Oxytelus nitidulus, Woll. Col. Atl. p. 517.

Hab. Azores (*Godman*); Europe; Madeira; Canaries.

Widely spread over Terceira, Fayal, Flores, and S. Miguel.

208. Trogophlœus riparius, Lac.

Trogophlœus riparius, Woll. Col. Atl. p. 518.

Hab. Azores (*Godman*); Europe; Madeira; Canaries.

Common in S. Miguel, Fayal, and Flores, in damp places.

209. Trogophlœus corticinus, Grav.

Trogophlœus corticinus, Woll. Col. Atl. p. 519.

Hab. Azores (*Godman*); Europe; Madeira; Canaries.

One specimen, from a stream near Horta, Fayal.

210. Trogophlœus subtilis, Er.

Hab. Azores (*Godman*); Europe.

Two specimens, with the preceding. The species is new to the Atlantic fauna.

211. Homalium pusillum, Grav.

Homalium pusillum, Woll. Col. Atl. p. 524.

Hab. Azores (*Godman*); Europe; Madeira; Canaries.

Two specimens only, from pines—one from Horta, and one from Santa Cruz, Flores.

212. Homalium clavicorne, Woll.

Homalium clavicorne, Woll. Col. Atl. p. 523.

Hab. Azores (*Godman*); Madeira.

Abundant in Euphorbia-stems in Flores; it is also not rare in Madeira.

List of Species.

	Europe, with N. Africa.	Madeira.	Canaries.	S. America.	Peculiar species.
1. Calosoma azoricum	..	..	*		
2. Blechrus maurus	*	*	*		
3. Licinus brevicollis	*				
4. Pristonychus complanatus	*	*	*		
5. Calathus flavipes	*				
†6. —— mollis	*				
7. Anchomenus aptinoides	..	..	..	..	*
8. —— albipes	*	*	*		
9. —— marginatus	*	*	*		
10. —— parumpunctatus	*				
11. Pterostichus nigerrimus	*	*			
12. —— vernalis	*				
13. Amara trivialis	*	*			
14. Anisodactylus binotatus	*	*			
15. Harpalus rotundicollis	*	*			
16. —— ruficornis	*				
17. —— griseus	*				
18. —— distinguendus	*	*			
19. Stenolophus teutonus	*	*	*		
20. —— brunneipes	*				
21. —— luridus	*				
†22. Bradycellus distinctus	*				
23. Trechichus fimicolus	..	*			
24. Tachys 4-signatus	*	*	*		
25. Bembidium rufescens	*	*			
26. —— schmidtii	..	*			
27. —— hesperus	..	..	..	..	*
†28. Hydroporus planus	*	..	*		
†29. Colymbetes pulverosus	*				
30. Agabus godmanni	..	..	..	..	*
31. Gyrinus dejeani	*	..	*		
32. Parnus prolifericornis	*	*	*		
†33. Philhydrus lividus	*				
†34. Sphæridium bipustulatum	*	*			
35. Dactylosternum abdominale	*	*	*		
†36. Cercyon obsoletum	*				
37. —— littorale	*	*	*		
38. —— inquinitum	..	*	*		
39. —— centromaculatum	*	*	*		
40. Corytholophus sublævipennis	*				
†41. Sericoderus lateralis	*	*	*		
†42. Ptenidium apicale	*	*	*		
†43. Phalacrus coruscus	*	..	*		
†44. —— consimilis	*	*	*		
†45. Carpophilus dimidiatus	*	*	*		
†46. —— mutilatus	*	*			
47. Meligethes incanus	*				
†48. Nitidula 4-pustulata	*	*			

	Europe, with N. Africa.	Madeira.	Canaries.	S. America.	Peculiar species.
†49. Nitidula colon	*	*			
†50. —— obsoleta	*	*			
†51. Monotoma 4-foveolata	*	*	*		
†52. —— spinicollis	*	*	*		
†53. —— quadricollis	*	*	*		
54. Tarphius wollastoni	..	..	..	..	*
†55. Aglenus brunneus	*	*	*		
56. Læmophlœus clavicollis	..	*	*		
†57. Silvanus advena	*	*	*		
†58. Nausibius dentatus	*	*	*		
†59. Cryptophagus cellaris	*	*	*		
†60. —— dentatus	*	*	*		
†61. —— affinis	*	*	*		
†62. —— punctipennis	*				
†63. —— saginatus	*	*			
†64. Cryptophorus schmidtii	*				
65. Paramecosoma simplex	..	*	*		
†66. Atomaria munda	*	*	*		
†67. Epistemus gyrinoïdes	*	*	*		
†68. Latridius minutus	*	*	*		
†69. —— nodifer	*				
70. Corticaria maculosa	..	*	*		
†71. —— fulva	*	*	*		
†72. —— serrata	*	*	*		
73. —— curta	*	*	*		
†74. Typhæa fumata	*	*	*		
†75. Mycetæa hirta	*	*			
†76. Dermestes frischii	*	..	*		
†77. Anthrenus varius	*	*	*		
†78. Acritus minutus	*	*	*		
†79. Carcinops pumilio	*	*	*		
†80. Saprinus cærulescens	*				
†81. —— semistriatus	*	*	*		
82. —— apricarius	*	*	*		
83. —— dimidiatus	*				
84. —— rugifrons	*				
†85. Onthophagus taurus	*				
†86. —— vacca	*				
†87. Aphodius granarius	*	*	*		
†88. —— lividus	*	*	*		
89. Psammodius sabulosus	*	*	*		
90. —— plicicollis	*				
91. —— porcicollis	*	*	*		
92. —— cæsus	*	*	*		
†93. Trox scaber	*	*			
94. Heteroderes azoricus	..	..	..	..	*
95. Æolus melliculus	..	..	..	*	
96. Monocrepidius posticus	..	..	..	*	
97. Elastrus dolosus	..	..	..	..	*
†98. Melanotus dichrous	*				

	Europe, with N. Africa.	Madeira.	Canaries.	S. America.	Peculiar species.
†99. Athous obsoletus	*				
100. Attalus miniatocollis	..	..	..	..	*
101. Malachius militaris	..	*			
102. Dolichosomus nobilis	*				
†103. Opilus mollis	*	*			
†104. Ptinus testaceus	*	*	*		
105. Mezium sulcatum	*	*	*		
†106. Anobium domesticum	*	*	*		
107. —— villosum	..	..	*		
‡108. —— paniceum	*	*	*		
†109. Ptilinus pectinicornis	*	*			
†110. Hylastes ater	*				
†111. Hylurgus ligniperda	*	*	*		
†112. Tomicus saxeseni	*	*	*		
†113. Hypoborus ficus	*	*			
114. Cryphalus aspericollis	..	*	*		
115. Mesites tardii	*				
116. Phlœophagus spadix	*	*			
117. —— tenax	..	*			
118. —— variabilis	..	..	..	..	*
†119. Calandra granaria	*	*	*		
†120. —— oryzæ	*	*	*		
†121. Ceuthorhynchus nigroterminatus	*	*	*		
122. Acalles droueti	..	..	..	..	*
123. Apion chalybeipenne	..	*	*		
†124. Pissodes notatus	*	*			
125. Laparocerus azoricus	..	..	..	..	*
†126. Otiorhynchus scabrosus	*				
†127. —— sulcatus	*				
†128. Hypera variabilis	*	*	*		
129. Asynonychus godmanni	..	..	..	..	*
130. Neocnemis occidentalis	..	..	..	..	*
†131. Sitona lineatus	*	*	*		
†132. —— flavescens	*				
†133. —— gressorius	*	*	*		
†134. Bruchus pisi	*	*	*		
135. —— tristiculus	*				
136. —— irresectus	*	*			
†137. Hylotrypes bajulus	*	*	*		
†138. Clytus 4-punctatus	*	*			
†139. Gracilia pygmæa	*	*	*		
140. Tæniotes scalaris	..	..	..	*	
†141. Leptura fontenayi	*				
†142. Haltica ampelophaga	*				
†143. Psylliodes chrysocephala	*	*			
144. —— vehemens	..	*	*		
†145. Coccinella 7-punctata	*	*	*		
146. —— 11-punctata	*				
147. —— variabilis	*				
†148. Chilocorus bipustulatus	*				

	Europe, with N. Africa.	Madeira.	Canaries.	S. America.	Peculiar species.
149. Scymnus durantæ	..	*			
†150. —— minimus	*	*	*		
151. Rhizobius litura	*	*	*		
152. Blaps gages	*	*	*		
†153. —— similis	*	*	*		
154. Hegeter tristis	*	*	*		
155. Opatrum hispidum	*	*	*		
156. Helops azoricus	..	..	..	..	*
157. Phaleria bimaculata	*	*			
158. Trachyscelis aphodioides	*	..	*		
†159. Tribolium ferrugineum	*	*	*		
†160. Tenebrio obscurus	*	*	*		
†161. Alphitobius piceus	*	*	*		
162. Anaspis? proteus	..	*	*		
†163. Anthicus floralis	*	*	*		
164. —— humilis	*	..	*		
165. —— hispidus	*	*	*		
†166. Falagria obscura	*	*	*		
†167. Aleochara nitida	*	*	*		
†168. —— puberula	*	*	*		
169. Homalota obliquepunctata	..	*			
170. —— luridipennis	*	*			
171. —— longula	*	*	*		
172. —— atricilla	*				
173. —— putrescens	..	..	*		
174. —— sp.?					
175. —— coriaria	*	*	*		
†176. —— nigra	*	..	*		
†177. —— atramentaria	*	*	*		
†178. —— melanaria	*	*	*		
179. Xenomma melanocephala	..	..	..	..	*
180. Habrocerus capillaricornis	*	*	*		
181. Conosomus sericeus	*	*	*		
182. Creophilus maxillosus	*	*	*		
183. Ocypus æthiops	*				
184. —— olens	*	..	*		
†185. Philonthus æneus	*	*			
†186. —— sordidus	*	*	*		
†187. —— umbratilis	*	*	*		
†188. —— scybalarius	*	*	*		
189. —— nigritulus	*	*	*		
190. —— filiformis	..	*			
191. Xantholinus glabratus	*				
†192. —— punctulatus	*	*	*		
193. —— hesperus	*	*	*		
194. —— linearis	*	*			
†195. Leptacinus pusillus	*	*	*		
†196. Stilicus affinis	*	*	*		
197. Sunius gracilis	*	*			
198. Lithocharis ripicola	*				

	Europe, with N. Africa.	Madeira.	Canaries.	S. America.	Peculiar species.
199. Lithocharis apicalis	*				
†200. —— ochracea	*	*	*		
201. —— ruficollis	*	*	*		
202. —— debilicornis	*	*	*		
203. Stenus guttula	*	*	*		
204. Platystethus spinosus	*	*	*		
†205. Oxytelus sculptus	*	*	*		
†206. —— complanatus	*	*	*		
†207. —— nitidulus	*	*	*		
208. Trogophlœus riparius	*	*	*		
209. —— corticinus	*	*	*		
210. —— subtilis	*				
†211. Homalium pusillum	*	*	*		
212. —— clavicorne	*	*			
Total	175	140	116	3	14

HYMENOPTERA AND LEPIDOPTERA.

Although the other orders of Insecta are all represented in the Azores, the materials I obtained are so incomplete that the enumeration of the species, even were it possible, would be of little scientific value. M. Drouet records a few insects of the orders Orthoptera, Hemiptera, and Diptera in his work already quoted, to which I refer my readers.

HYMENOPTERA.

I brought home but a small number of insects belonging to this order, as the following list will show. For the determination of the species I am indebted to Mr. Frederick Smith of the British Museum.

Fam. ICHUEUMONIDÆ.

1. ICHNEUMON ANTENATORIUS, Panzer.

Ichneumon nigerrimus.

Hab. Azores (*Godman*); Europe.

Widely distributed, probably introduced into the Azores.

Fam. FORMICIDÆ.

2. FORMICA EMARGINATA.

Hab. Azores (*Godman*).

Common in S. Miguel.

3. PHEIDOLE PUSILLA, Heer.

Hab. Azores (*Godman*); Madeira.

4. TAPINOMA COLLINA, Smith.

Hab. Azores (*Godman*); Madeira.

5. MYRMICA CARBONARIA, Smith.

Hab. Azores (*Godman*); Madeira.

This insect has not, I believe, been yet found beyond the above islands.

6. MYRMICA ——?

Hab. Azores (*Godman*).

Of this species I unfortunately obtained, in S. Miguel, specimens of one sex only. Mr. Smith, however, tells me that it is probably new.

Fam. VESPIDÆ.

7. VESPA VULGARIS, Linn.

Vespa vulgaris, Drouet, Faun. Açor. p. 203.

Hab. Azores (*Godman, Drouet*); Europe.

I found this insect in all the islands I visited, but nowhere very abundant.

Fam. Apidæ.

8. Osmia emarginaria, St. Farg.
Hab. Azores (*Godman*); S. Miguel and Terceira; Europe.

9. Apis mellifica, Linn.
Apis mellifica, Drouet, Faun. Açor. p. 203.
Hab. Azores (*Godman, Drouet*); Cosmopolitan.

10. Bombus ruderatus, Fabr.
Hab. Azores (*Godman*); Madeira.
Is closely allied to *B. hortorum,* a very common European species. Common in all the islands I visited.

11. Halticus cylindricus, Fabr.
Hab. Azores (*Godman*); Europe.

12. †Anthidium manicatum, Linn.
Anthidium manicatum, Drouet, Faun. Açor. p. 203.
Hab. Azores (*Drouet*); Europe.
This is not amongst the few insects I brought, but is inserted on M. Drouet's authority.

13. Megachile centuncularis, Linn.
Hab. Azores (*Godman*); Europe.

LEPIDOPTERA.

The Azores are exceeding poor in Lepidoptera, not only in the number of species but also in individuals. With one exception, however, the *Rhopalocera* are all common European insects, as the following list will show. For the names of the *Heterocera* I am indebted to Mr. H. T. Stainton.

Rhopalocera.

1. Danais archippus, F.
Hab. Azores (*Godman*); North and Central America.

I met with only two specimens of this insect, neither of which did I catch myself. One was taken in Flores in 1864, the other kindly given me by Mr. J. Dabney of Fayal, where it had been caught the previous summer (1864). Both specimens are females. I have compared these specimens with others I have in my collection from North America, with which they perfectly agree. I do not regard the species as established in the Azores, though the fact of its having been obtained from two islands so widely separated is a curious coincidence, and not easily accounted for. I met with no one who knew the insect or had ever seen it before. I may also add that there is regular communication between North America and the Azores.

2. Epinephele janira, Linn.

Satyrus janira, Drouet, Faun. Açor. p. 204; Morel. Hist. Nat. des Açor. p. 96.

Hab. Mountains of S. Miguel and Santa Maria (*Drouet*); Europe.

I did not meet with this species myself, and therefore insert it on MM. Morelet and Drouet's authority.

3. Pyrameis cardui, Linn.

Vanessa cardui, Drouet, Faun. Açor. p. 204; Morelet, His. Nat. des Açor. p. 96.

Hab. Azores, eastern and central groups (*Godman*); Europe, &c.

Tolerably common. In Madeira the American *V. hunteri* is found.

4. Pyrameis atalanta, Linn.

Vanessa atalanta, Drouet, Faun. Açor. p. 204; Morelet, Hist. Nat. des Açor. p. 96.

Hab Azores (*Godman*); Europe.

Found in all the islands I visited.

5. Pieris brassicæ, Linn.

Pieris brassicæ, Drouet, Faun. Açor. p. 203; Morelet, Hist. Nat. des Açor. p. 96.

Hab. Azores, eastern, central, and western groups (*Godman*); Europe.

6. Pieris daplidice, Linn.

Pieris daplidice, Drouet, Faun. Açor. p. 204; Morel. Hist. Nat. des Açor. p. 96.

Hab. S. Miguel (*Drouet*); Europe.

7. Pieris napi, Linn.

Pieris napi, Drouet, Faun. Açor. p. 204; Morel. Hist. Nat. des Açor. p. 96.

Hab. S. Miguel (*Drouet*); Europe.

8. Pieris rapæ, Linn.

Pieris rapæ, Drouet, Faun. Açor. p. 204; Morel. Hist. Nat. des Açor. p. 96.

Hab. S. Miguel (*Drouet*); Europe.

As in the case of *Epinephele janira*, the three Pieridæ last mentioned are inserted on MM. Morelet and Drouet's authority. I did not meet with any of them myself.

9. Colias edusa, Fabr.

Hab. Azores, eastern and central groups (*Godman*); Europe.

I met with this wide-ranging species in some numbers both in S. Miguel and Fayal.

Heterocera.

10. Macroglossa stellarum, Linn.

Macroglossa stellarum, Drouet, Faun. Açor. p. 204; Morel. Hist. Nat. des. Açor. p. 96.

Hab. Pico and Fayal (*Godman*); Europe.

Tolerably common.

11. Deilephila nerii.

Deilephila nerii, Drouet, Faun. Açor. p. 204; Morel. Hist. Nat. des Açor. p. 96.

Hab. Fayal (*Drouet*); Europe.

M. Drouet says this is a common species; but I did not see it.

12. Sphinx convolvuli, Linn.

Sphinx convolvuli, Drouet, Faun. Açor. p. 204; Morel. Hist. Nat. des Açor. p. 96.

Hab. Fayal (*Godman*); Europe.

I procured two or three individuals only from this island.

13. Sphinx ligustri, Linn.

Sphinx ligustri, Drouet, Faun. Açor. p. 204; Morel. Hist. Nat. des Açor. p. 96.

Hab. Fayal (*Drouet*); Europe.

I did not observe this species myself.

14. Acherontia atropos, Linn.

Acherontia atropos, Drouet, Faun. Açor. p. 204; Morel. Hist. Nat. des Açor. p. 96.

Hab. S. Miguel, Fayal, Pico (*Godman*); Europe.

I saw several examples of this widely distributed insect.

15. Leucania extranea.

Hab. S. Miguel (*Godman*); Madeira.

[This species has been taken twice in Great Britain, but is not yet known to occur in Continental Europe, though found in most other parts of the globe.—H. T. S.]

16. Agrotis segetum, Wien. Verz.

Hab. S. Miguel (*Godman*); Europe.

17. Agrotis saucia.

Hab. Fayal and S. Miguel (*Godman*); Europe.

[This species has a wide geographical range.—H. T. S.]

18. Plusia gamma, Linn.

Hab. Fayal (*Godman*); Europe.

19. Hypena obstitialis.

Hab. Fayal and St. Michael's (*Godman*); Madeira; S. Europe.

20. Geometra ——?

Hab. Terceira (*Godman*).

21. Geometra ——?

Hab. Fayal (*Godman*).

22. Camptogramma fluviata.

Hab. Terceira and Fayal (*Godman*); Europe.

[Formerly a great rarity in England, but now common enough. It has also a wide range.—H. T. S.]

23. Phibalapteryx polygrammata?

Hab. Flores (*Godman*); Madeira.

[This species is interesting, as it may be specifically distinct from our European insect.—H. T. S.]

24. Pyralis farinalis, Linn.

Hab. Fayal; Madeira (*Godman*); Europe.

[Probably introduced through trade.—H. T. S.]

25. Botys ferruginalis.

Hab. Fayal (*Godman*); Madeira.

26. Eudorea, n. sp.?

Hab. Fayal and Flores (*Godman*).

27. CARPOCAPSA POMONELLA, Linn.
Hab. Terceira (*Godman*); Europe.
[No doubt introduced through trade.—H. T. S.]

28. DEPRESSARIA —— ?
Hab. Flores (*Godman*).

TERRESTRIAL MOLLUSKS OF THE AZORES.

By the Rev. H. B. TRISTRAM, M.A., F.R.S., &c.

THE Pulmonifera of the Azores have been the subject of special attention, and had been thoroughly worked out before any other branch of the fauna of the group had attracted attention. This was due to the researches of Morelet and Drouet. Yet the result of very careful explorations reveals rather the poverty than the richness of the land, as might have been anticipated on a volcanic and comparatively recent soil. It is on limestones and chalk that the greatest number both of individuals and species are generally found; and volcanic regions are even more destitute than granitic of pulmoniferous Gasteropods. Thus, to compare the Azores with the other Atlantic island groups, there are more than 120 species known from the Madeiras, 105 from the imperfectly explored Canaries, and only 69 from the Azores. These, again, are for the most part inconspicuous and minute, many of the *Pupa* tribe even microscopic; and none can compare for an instant in size, colour, or beauty with those of the other islands. We are speaking here especially of those species which are *peculiar* to the Western Isles.

But the mollusks reveal very strongly European affinities, and unite the Azores more closely with the continent than with the other Atlantic islands. Thus of the 69 Azorean species, while 4 (or 6 per cent.) are identical with Canarian, and 7 (or 10 per cent.) with Madeiran species, no less than 26 (or 38 per cent.) are of Western Europe, including 10 which are found generally in the Atlantic zone. Of the 32 species (46 per cent.) peculiar to the Azores, none attain any size; most are inconspicuous and minute; and I conceive that 27 out of the 32 would be admitted to be more closely allied to common European types than to any others, thus leaving only 5 (or 7 per cent.) of the whole number which can be considered strikingly peculiar. Of the *Limacidæ* Morelet has noted 9 species, 3 *Arion*, 4 *Limaces*, 1 *Testacella* (*T. maugei*), all European, and one of a new genus, the most interesting and peculiar gasteropod discovered in the Azores, *Viquesnelia atlantica*, Morelet. Its affinities are between *Limax* and *Parmacella*; and it has no congeners either in the European or Atlantic zone. The only other living species is Indian; and fossil remains of a similar mollusk have been found in the nummulitic limestone of the Pyrenees, and in Roumelia. This slug is of a reddish-brown colour, its skin wrinkled and dry, and a large protuberance on the back marks the shield and covers the internal rudimentary shell. Unfortunately this singular and interesting creature did not come under Mr. Godman's personal observation.

Of the *Helicidæ* the most characteristic genus in the Azores is *Vitrina*, of which no less than 7 species have been described—*Vitrina pelagica*, *V. laxata*, *V. brumalis*, *V. mollis*, *V. brevispira*, *V. finitima*, and *V. angulosa*. Of these, *V. laxata* is the only one which attains any size, none of the others exceeding much the European kinds, and *V. angulosa* being minute. They were all first described by Morelet. Different species pertain to different islands of the group; but four species are found on the island of Santa

Maria. *V. laxata* recalls the type of the magnificent members of this genus found in the Madeiras, while *V. finitima* exhibits a remarkable reflection of the peristome, which distinguishes it from all others. For our knowledge of this group we are dependent on MM. Morelet and Drouet.

In the genus *Zonites*, Mr. Godman has been more fortunate than in the preceding, having collected several species in considerable numbers. Six species have been recognized, of which by far the most abundant is the universal *Z. cellarius*, which attains a size considerably larger than in Northern Europe. The next plentiful species is *Z. atlanticus*, Morelet, a pretty little shell, very distinct from any species found elsewhere, but collected by Mr. Godman from all the islands, remarkable for its imperforate character, in which respect it closely resembles *Helicella suppressa*, Say, of North America. It is worthy of remark that all the peculiar forms of *Zonites* resemble very closely the North-American types. Larger than *Z. atlanticus*, but similarly imperforate and transversely striped by yellowish rays, is *Z. miguelinus*, Pfeiff. The third species peculiar to the Azores is *Z. volutella*, Pfr., a remarkably elegant, discoidal, deeply umbilicated shell, brightly striped transversely with yellowish-red lines. It has been well figured by Morelet, and was found by Mr. Godman in Fayal. The other members of the genus occurring are *Z. crystallinus*, Müll., and *Z. fulvus*, Müll., both common European and British species, but rather scarce in the Azores.

Of the more restricted genus *Helix* we find 21 species, of which 8 are peculiar to the islands. Nine species are European, viz. *H. pisana*, the most abundant of all, *H. lactea*, *H. aspersa*, *H. barbula*, Charp., so common in Northern Spain and in Portugal, and abounding everywhere in the Azores, *H. lenticula*, *H. rotundata*, *H. acu-*

leata, *H. pulchella*, and *H. apicina*. Most of these occur in Mr. Godman's collection. Of species belonging to the Madeiras, Mr. Godman has collected *H. armillata*, Lowe, and *H. erubescens*, Lowe. *H. paupercula*, Lowe, is also very common in Pico and Fayal. One minute shell, *H. servilis*, Shuttl., has been found by Morelet in the Azores, which had only hitherto been recognized from the Canaries. Its extreme minuteness has probably prevented its being noticed as yet in the Madeiras.

The remaining *Helices* are all exclusively Azorean. Of these only one has been found by Mr. Godman, *H. horripila*, Morelet, a curious little shell, stated by Morelet to be very scarce. It is covered with fine spines or hairs very closely set, something like *H. setosa*, and delicately striated. The umbilical perforation is deep and contracted. Another species is *H. azorica*, Albers, a very fragile shell, imperforate, prettily marked, and rather recalling some of the Madeiran species. To the same group belong:—*H. terceirana*, Morelet, a larger shell, solid, and resembling *H. arbustorum*; *H. caldeirarum*, Morelet, a scarce species inhabiting San Miguel; *H. drouetiana*, Morelet, still more rare, only found in one or two spots on Fayal. A species allied to this, *H. niphas*, Pfr., is described by Pfeiffer as from San Miguel, but has not been obtained by Morelet, Drouet, or Godman, nor am I aware of the existence of a figure. The other species are minute:—*H. vespertina*, Morel., found in Terceira, scarce; *H. monas*, Morel., found in San Miguel, and also very rare.

There are two other species (*H. vetusta*, Morelet, and *H. obruta*, Morel.) which have as yet been only found fossil, in company with fossils of other existing species, both at Santa Maria,—the former resembling closely some Algerian species; the latter an inconspicuous shell, not far removed from *H. armillata*, but larger.

The genus *Bulimus* is well represented, by 9 species. Two of these are common European forms, *B. decollatus* and *B. ventrosus,* found in some plenty by Mr. Godman. *B. variatus,* Webb & Berth., is the only other non-peculiar species, having been first described from the Canaries. In the Azores it occurs sparsely in Santa Maria.

The six special species are :—*B. pruninus,* Gould, the most abundant and variable shell in the whole islands. It is a very distinct species, belonging to the common European type, but larger and more solid than any of its congeners. It is also remarkable for its bright and variable colouring, which ranges from dark purple, the ordinary hue, to light pink and even white. The peristome is very thick, and reflected, of a pearly white. The other five species present nothing remarkable in their form or coloration; they are all nearly allied to the plain-coloured group, exemplified in our *B. lackhamensis.* They are :—*B. forbesianus,* Morel., = *atlanticus,* Forbes; rather scarce, but widely distributed. *B. vulgaris,* Morel.; found in some plenty by Mr. Godman in Fayal and San Miguel. *B. delibutus,* Morelet; Fayal and Terceira. *B. hartungi,* Morelet; Island of Santa Maria. *B. santa-marianus,* Morelet, very much more oval and globose than the other species, and only found on the island from which its name is derived.

The family *Achatinidæ* are only represented by the common European *Zua lubrica,* found by Mr. Godman in various places.

The Pupas are numerous in species but sparse in individuals, and all very minute. *Balæa perversa,* identical with our British mollusk, is found on all the islands. *Pupa pygmæa* has been noticed, but rarely, by Morelet in San Miguel. The same author also records two Madeiran species—*Pupa microspora,* Lowe, and *P. anconostoma,*

Lowe, of which the former seems to be very scarce. The latter, very closely allied to *P. umbilicata* of Britain, and very possibly only a stunted variety of it, is very common, and has been found also by Mr. Godman.

Five other minute (in fact, microscopic) species are described for the first time, and admirably figured on an enlarged scale by Morelet. They all belong to the same group as our minute British Pupas, now separated as *Vertigo*, and are *P. tessellata*, *P. fasciolata*, *P. fuscidula*, *P. rugulosa*, and *P. vermiculosa*. *P. fasciolata*, the only one at all common, was obtained also by Mr. Godman. The others are scarce; and *P. rugulosa* depends on a single specimen found in the island of Pico.

Three species of *Auricula* or *Conovulus* are found in Pico on the sea-washed rocks—*Auricula vulcani*, Morel., and *A. vespertina*, Morel., both also Canarian, and *A. bicolor*, Morel., extremely like our *A. bidentata*. *Pedipes afer*, Fér., another Canarian species, is also found in their society.

Of the operculated Pulmonifera, the Azores present but two species, and those decidedly not European in their affinities. They are the small *Cyclostoma hespericum*, Morelet, found by him on most of the islands, and the minute *Hydrocena gutta*, Shuttl., which is scarce. This latter is also found in the Canaries. The *C. hespericum*, though peculiar, is manifestly related to the Madeiran species of the group.

It should be observed that, of all the Pulmonifera of the Azores, *Pedipes afer* is the only one common to the African continent.

The class of Gasteropods is by far the most numerous of all the forms of life in the Azores; and among them are found a larger proportion of peculiar species than in any other class. But the affinities of all are more decidedly

European than African. There is no trace of American influences—except in the general resemblance of some members of the genus *Zonites*; even of these the coast of Portugal seems to have supplied the greater proportion. The intrusion of so many Madeiran and of a few Canarian species is what might have been anticipated from the geographical position of the Azores.

There is, however, one singular hiatus in the molluscan fauna. Though there are abundant streams, springs, and lakes presenting the most favourable conditions for their existence, not a single representative of the Pulmobranchiate Mollusca has yet been discovered. These are to be found in every other portion of the globe. Not an island in the Pacific, not even Greenland and Iceland, which are beyond the usual range of Pulmonifera, are without representatives of this class; yet in the Azores no species of the world-wide genera of *Limnæa, Physa, Ancylus, Neritina, Cyclas,* or *Cyræna* has yet been found. Perhaps one cause of this deficiency may be due to the small number of straggling waders and ducks, which birds seem often to be the means of the dispersion of fluviatile mollusks.

BOTANY OF THE AZORES.

By H. C. Watson.

I. BOTANICAL LITERATURE, &c.

Passing over the earlier records of individual plants, chiefly collected by Masson, the first Flora Azorica bears the date of 1844. Dr. Mauritz Seubert was the author, and also the illustrator by his own pencil, of that valuable contribution to our stock of Local Floras. His work was based chiefly on collections made in the Azore Isles by C. F. Hochstetter in 1838. By a reference given in the preface to the Flora, it appears that a list of the plants had been published the preceding year in Wiegmann's Archiv für Naturgeschichte, ix. Some few other species are also taken into the Flora from books previously published.

Dr. Seubert's volume still remains the only real Flora of the Isles; it is a full enumeration of the species, so far as known to the author, with occasional remarks, diagnostic characters of the new species being introduced, and several of them being illustrated by characteristic plates. The chief defect of that Flora now is, that later researches have rendered it a very incomplete enumeration of the plants actually found in the Isles; so that a copious supplement would now be needful to bring it up to presently existent knowledge of Azore botany.

Some minor defects may also be mentioned by way of

caution. Dr. Seubert was placed under the inconvenience and great disadvantage of writing the Flora of a country which he had not seen. Thus his work is truly more a botanical account of dried specimens from the Azore Isles than a proper Flora of those isles; and perhaps it would have been better had he even more strictly limited himself to such an account, avoiding guesses that might prove only erroneous records. For instance, he gives alleged ranges of altitudes at which various of the species are stated to occur; but it is asserted here with some confidence that the stated altitudes must too often have been merely rough guesses by somebody not sufficiently informed about the true heights of hills and places in the Isles. Further, it too frequently happens that the habitats are given in loose and vague terms, not to say in false terms. Instead of naming the islands in which the species had been observed, or in which the specimens under view had been picked, we see continually such loose expressions as "hinc inde," "plurium insularum," "omnium insularum," "cum prioribus," "cum precedente," &c.; while some of these loose descriptions of "Habitat" are given for species which, there is too much reason to fear, were not found at all, seeing that the other collectors, presently to be named, one of them a regular resident, have failed to find the plants. Examples of this kind will be given in the new Catalogue subjoined. What is the proper meaning of "omnium insularum" is not explained, while it cannot be literally or arithmetically true. We may guess that it means all the six islands (out of nine in the whole) actually visited by Mr. Hochstetter. But in such reading we should be always including the little isle of Corvo for species which too likely were not seen there; to say nothing about other islands also, in which they may or may not have been found. Still, these are minor matters. As a whole, the Flora Azorica is a good work, creditable to its author, and serviceable to progressive science.

During four summer months of 1842, May to September, I was allowed to occupy a cabin on board the British war-steamer 'Styx,' then engaged in the peaceful duty of surveying the Isles, by direction of the British government. My object in this was purely botanical investigation, undertaken at the suggestion of Sir William Hooker. Unfortunately the plan was not found so favourable as expected for the object specially in my view. That summer the steamer visited four only of the nine islands; these four not including the two principal islands of the group, San Miguel (or Saint Michael's) and Terceira. Moreover, very insufficient facilities were afforded me for passing to and fro between the ship and the shore. And when I took advantage of the boats of the island fishermen for this purpose, I lost much of the day through delays in the morning and through feeling compelled to be back on board the English vessel at night, lest she should have gone elsewhere, leaving me behind. Her Captain (since deceased) was an able surveyor, an intelligent and agreeable man as an acquaintance on shore. But among other somewhat strange ideas of ship-discipline, was a notion that he best secured the readiness of his officers and crew by keeping everybody in ignorance of his intentions, or the intended positions and movements of his ship from day to day. For my pursuits this proved a very inconvenient practice; an expensive matter also in the frequent hire of boats for conveyance from shore to ship, sometimes to the vessel miles away from the land. Besides which, it subjected me to the great inconvenience of drying my plants in the spare space of a small sleeping cabin, some six feet by five in its dimensions. In consequence, at first I limited myself to saving only three specimens of a species; and although this scanty numerical ratio was afterwards exceeded for most of the species having any special interest, yet the total collection remained much too small; and after my return home the duplicates were too hastily distributed to botanical friends; leaving only the few examples reserved for my own herbarium; the insufficiency of which afterwards partly led me into some errors in their specific names. Eventually, this deficiency was greatly remedied in a way that will presently be explained.

In the second volume of the London Journal of Botany, by the desire of Sir William Hooker, I wrote three brief papers giving my first impressions of Azorean botany, as felt by a young botanist, whose previous experience had been limited to the plants of more northern latitudes, and who was then scarcely at all familiar with the plants even of South Europe, with which those of the Azore Isles have in general the closest affinity. The flora of the Isles, indeed, is a scanty fragment of the Mediterranean flora, reinforced by a small percentage of non-European species, partly peculiar to the Isles, partly shared by them with Madeira or the Canaries.

Subsequently, in the third volume of the same Journal, I printed a general list of the species collected by myself; incorporating therewith also the species enumerated in the Flora Azorica, in order to make up a more complete list of the Azore plants than either would have been if kept apart from the other. For some years, therefore, this list remained the most complete one in print, though without at all pretending to supersede the true Flora Azorica, the work of Dr. Seubert. Again, in the sixth volume of the same Journal of Botany, pp. 380–397, I added a Supplementary List and Notes, increasing the recorded flora of the Isles by about fifty additional species, brought to my knowledge by a resident botanist in Saint

Michael's, Thomas Carew Hunt, British Consul for the Azores.

The zealous and intelligent exertions of Mr. Hunt in the years 1844–48 conduced much to the increase and to the diffusion of knowledge about the botany of the Isles. He most kindly collected for me, and for the (then flourishing, since extinct) Botanical Society of London, a very large supply of Azore plants, chiefly from the island of his own residence, San Miguel; thus excellently supplementing my own unsatisfactory collection by the number and goodness of his specimens, and by the fresh locality from which they were procured. These specimens were promptly distributed into British, European, and American herbaria. Their labels were printed in London, and filled in by my own pen. All bear the same numbers on them, and the same names, with few exceptions, as had been used on my own earlier labels. Mr. Hunt's added species of course would have no corresponding number; they usually bear the abbreviation "Add." instead. Thus the two collections united, and bearing the same manuscript numbers or substituted abbreviation on their labels, have been rendered in some sort published records of Azoric botany. From time to time, however, some few corrections of name were made, so that the later-dated labels, in or about 1848, are more to be trusted than my own earlier labels of 1842, or those with Mr. Hunt's specimens in 1844 or 1845. Since the publication of the Supplementary List several other additional species have been discovered and sent by Mr. Hunt, again further extending the known flora of the Isles.

Mr. Hunt has himself also published some botanical account of the Isles in his description of St. Michael's and St. Mary's, in the Journal of the Geographical Society of London for the year 1845,—a serial which has never come under my own view. The fact is stated in the work of

M. Drouet, presently to be mentioned. Mr. Hunt rather abruptly ceased correspondence with me, very much to my own regret, and on what grounds I remain entirely ignorant. Lately I have been informed that he made a wide change of residence and official duties from Saint Michael's to Stockholm. I am not aware that he continued his investigations into Azore botany after 1848 or 1849, if he remained at Saint Michael's to any later date.

Besides the collections of wild specimens above mentioned, a considerable number of garden-grown examples of the more peculiar or doubtful species have also been distributed by myself at various subsequent dates. These examples were grown in my own garden in the county of Surrey, partly from seeds brought home by myself, partly from seeds transmitted by Mr. Hunt. Altogether many thousands of wild or cultivated specimens have been thus distributed into herbaria, from which of course I have enjoyed the advantage of selection for my own herbarium. The watching of those garden-grown plants, some of them repeatedly raised by seed-sowing during a quarter of a century, ought to have sensibly aided me in attaining more correct conclusions regarding their specific affinities and diversities, when compared with British and European plants.

In 1866 a new Flora of the Azores was published under title of 'Catalogue de la Flore des Iles Açores,' from the pen of Henri Drouet, favourably known as a zoologist. M. Drouet, along with his companion M. Arthur Morelet, had visited the Isles in 1857, primarily with the purpose of zoological investigation, but the two devoting considerable attention also to botany. At the same time likewise Mr. Hartung, a German geologist, travelled in the Isles, uniting botany with his own specialty. The new Flore or Catalogue by M. Drouet is an industrious compilation, combining into one list all the plants recorded by his predecessors, together with those observed by himself or his two scientific fellow travellers. In Drouet's Catalogue the species actually seen by himself, or by either of the two other travellers, are judiciously distinguished by a mark affixed to their names in the list of species. I shall presently have frequent occasion to mention the fact of a mark or of no mark to the names, in commenting on some of the species more or less doubted as inhabitants of the Isles.

In some respects M. Drouet has been over sedulous to include in his enumeration the name of every thing reported as Azoric. His own chief specialty being zoological, he was not sufficiently familiar with the synonyms and name-changes of species in botany to escape committing sundry errors. This is particularly evident in the many species which are enumerated twice over in the catalogue, under different specific names, and occasionally also under different generic names. The number of Azore plants is thus unduly augmented, ostensibly amounting to 600 species, exclusive of *Cellulares,* against Seubert's original 325 species or thereabouts, which have been since gradually increased to nearly 500, but including among these 500 a considerable number of species introduced through gardens or field crops, and only casual or adventitious intruders among the proper Azoric plants.

In his prefatory account of botanical literature relating to the Azore Isles, Drouet mentions a Catalogue of plants in the Botanic Garden of the School of Medicine in Lisbon, in which many Azore plants are indicated. He states that the catalogue named was carefully drawn up by "MM. B. A. Gomès and Da S. Beirâo." That catalogue has not come under my own inspection; any thing from it incorporated with the observations made by M. Drouet and fellow travellers will appear at second-hand in the subjoined list of species and their insular habitats. M. Drouet usually names the islands in detail, but not seldom the expression "tout l'archipel" is used instead. Now, of the nine islands in the whole, two are very small, and these two appear to possess scanty floras, especially Corvo. Two others apparently have been little examined by botanists, San Jorge (St. George) and Graciosa. Thus, the expression "tout l'archipel" in Drouet's Catalogue, like that of "omnium insularum" in the Flora Azorica, may too often mean for certainty only the other five islands, with or without the occasional addition of one or more of the remaining four.

The circumstances under which Mr. Godman's collection of plants was made in 1865 are explained by himself. The collection was kindly placed before me by Mr. Godman, with the liberal permission to select any specimens serviceable for my own herbarium. As I rapidly turned over the plants advantage was taken of that permission in the case of any present doubt arising, or of any point reserved for after consideration occurring to me. The collection had previously been placed at the selective disposal of the Herbarium-keepers at Kew; and a manuscript list of the whole made it very evident that an ample selection had been taken out (perhaps, rather, that a full set of them had been taken, without any selection at all), several of the species named in the Kew list being no longer represented by any specimens in the collection; and as I found much ground for suspecting that the hastily made out list (not pretending to be a critically exact one) included some errors of name, I went over to Kew in the hope of seeing the specimens collectively before their dispersal through the jumble of riches and rubbish which makes up that vast and valuable herbarium. Unfortunately for my object the specimens had been already much dispersed,—some away in the hands of the gluers, some gone into their arranged places in the herbarium, &c. Professor Oliver, however, most kindly and attentively met my wishes, so that through his aid I was enabled to see considerably the larger portion of Mr. Godman's specimens, although the time at command allowed only a very cursory glance at them. Mingled with the specimens given by Mr. Godman, I found also some sent by the Barão do Castello de Paiva, to which occasional reference will be made in the new catalogue of species subjoined.

These explanations are thus entered upon because the total collection made by Mr. Godman was so good, both in respect of the number of the species and of the condition of the specimens, that I have felt it to be proper and desirable to cite that collection regularly in the subjoined catalogue; and as examples of nearly all of the species probably now (1868) are or will be placed in the Kew Herbarium, the abbreviated form of "Godman coll." becomes a reference to them there. In most instances, where I failed to see an actual specimen of the alleged species, the citation is changed into the words "Kew list of Godman coll." This will mean that I rely solely on the manuscript list above mentioned for the fact that the species under consideration was represented in the collection. But the selected specimens being in disarray at the time, as above explained, and my own time for looking them over being very restricted, I may not always correctly adhere to the distinctive form of citation.

On the request of Mr. Godman, and from the various records and materials thus explained, I now endeavour to draw out a new catalogue of Azore plants, which shall be either more complete or more correct than the preceding lists. Under the circumstances the subjoined catalogue will necessarily assume a sort of critical or criticising character. It will be a compilation from the previous writings of myself and of others on the same subject, with my own corrected views subsequently formed by further examination of the plants or specimens themselves. If those views should occasionally appear to be expressed in terms too positive or dogmatic, it must be kept in mind that I state them usually after inspection of a large stock of specimens, and from a fair store of experience through observing many of the plants living, as well as the dried examples of them.

A few words on the plan or method adopted, which is simple enough. It has generally been preferred to keep to the names under which the species have previously been recorded as Azore plants, not changing to newer names for the same species, unless some strong reason appeared to render the innovation more desirable than the uniformity. Usually, in case of the same species appearing under different names in the works cited, the diversity of nomenclature is made apparent by special mention of it, or by quotation in some form. In this way it is sought to render the new Catalogue a sort of general index to all the preceding lists.

Following the name of the species, some indication of its habitat outside the Azore Islands is added. The immediate or primary affinity of the Azore flora is with that of south and west Europe, and the word "Europe," with some fractional addition or substitute, will show the species so shared in common. Failing a known habitat in Europe, then Madeira and the Canaries will be cited, one or both according to circumstances. This will indicate the more special geographical affinities of nearly all those species which appear to be non-European, without being limited exclusively to the Azore Islands themselves. But the published lists of plants for Madeira and Canaries, neither of them, can be considered complete or clear from errors, and the citations of both habitats may occasionally prove corroborative of each other. Failing Europe, Madeira, and Canaries, any other country will be named as a habitat, and these will be few instances. The Manual Flora of Madeira, by Mr. Lowe, is much the best authority for that island, so far as yet published, Ranunculaceæ to Compositæ in part. Webb and Berthelot's more ambitious work on the Canaries, although a valuable contribution to the literature of natural science, is perhaps not equally reliable in its botanical details. True, I may consider Mr. Lowe too prone to rely upon petty and inconstant technical distinctions, but such a bias is one towards the safer side in a very local Flora. If the species under treatment is an accepted native, or a well established colonist in our own islands, England or Britain may be the fractional part of Europe to be specially named.

Next comes the number of Azore Islands in which I find the species stated to occur, taking printed records and manuscript labels into account. The islands themselves are then severally named; and afterwards the five authorities before mentioned, or such of them as have recorded the species in books or by labels with specimens, omitting them when they simply repeat the records of their predecessors already cited and without confirmation afresh. Thus, if a species in Drouet's Catalogue bears affixed to its name the mark of having been actually seen by M. Drouet or his fellow travellers, in such case "Drouet cat." is cited; but where that work simply repeats the record of a predecessor, the citation of it is not made, unless under some special consideration which renders a citation of it either needful or serviceable. Remarks on the nomenclature and technical characters of the species under treatment, with various miscellaneous observations, are added as the occasion may require or appear to warrant. The new Catalogue will thus be made a key or index to all the earlier-dated floral lists for the Isles; and likewise it will itself become a more true list of the presently known species, approximately complete and correct for the time being, although doubtless further additions and corrections will be made in the future.

II. CATALOGUE OF PLANTS.

1. Ranunculaceæ.

1. Ranunculus flammula, Linn. Europe. Britain. Isle 1. Terceira; Morelet, in Drouet flore, no. 1, the only collector who appears to have met with this common European plant. Is there any mistake concerning its existence in Terceira? Absent from Madeira and Canaries, it is reported to occur in Spain and Algeria.

2. Ranunculus grandifolius, Lowe. Madeira. Isles 4. San Jorge, Fayal, Pico, Flores. Seubert flo. 305. 48. Watson cat. 1. Hunt coll. Kew list of Godman coll. Drouet cat. 5.

Under name of *Ranunculus cortusæfolius* (of Willdenow) in the Flora of Seubert, as also in the other Catalogues cited. But the plants of Madeira and the Azores appear sufficiently different from the true *cortusæfolius* to be easily accepted for distinct species. Whether that of the Azore Isles is properly united with the species of Madeira, may admit of further question. The three Azore examples still retained in my herbarium (from Fayal and San Jorge) are remarkable by their copious and coarse pubescence, and by the blunt and broad teeth of their very thick leaves, characters by which they diverge from the Madeiran specimens in the same herbarium. I have seen the fruits in a young stage only.

3. Ranunculus repens, Linn. Europe. Britain.
Isles 3, or more. Miguel, Pico, Fayal. Seubert flo. 304. Watson cat. 2. Hunt coll.

Whether this widely distributed species is truly native in the Isles, or an established alien, may be questionable. It occurs under two conditions,—as an ordinary weed by road-sides and in cultivated ground, but likewise occasionally amid the grass-sward of the lower hills. The manner in which it stands recorded in the Flora of Seubert, and in the Catalogue of Drouet, may be commented upon here in order to exemplify a difficulty much too frequently experienced in quoting from those works. No islands are named as habitats for this plant in either book. By Seubert it is reported "hinc inde in cultis et ad vias;" and apparently (the usual label number being omitted) no specimens were collected by Hochstetter for distribution. In turn, by Drouet it is said to inhabit "tout l'archipel;" while it appears, nevertheless, not to have been observed at all by himself or his fellow travellers, as the name stands in his book without the mark which he uses to distinguish the species actually seen by them. How M. Drouet got the knowledge which entitles him to state that the plant occurs in all the nine islands, thus seems difficult to make out. The fair inference from this, and from various other such instances, would seem to be that the general expressions, more or less positively implying "all the islands," were introduced at random or by sheer guess into either work, and more especially into that of M. Drouet.

4. Ranunculus trilobus, Desf. South Europe.

Isles 4. Miguel, Maria, Terceira, Flores. Seubert flo. 304. 47. Watson cat. 3. Hunt coll. Drouet cat. 3.

5. Ranunculus parviflorus, Linn. Europe. Britain.

Isles 5. Miguel, Maria, Terceira, Pico, Fayal. Seubert flo. 308. 46. Watson cat. 5. Hunt coll. Kew list of Godman coll. Drouet cat. 4.

Seubert enumerates also a variety of this species under the name of "*acutilobus*," no. 308*a*.

6. Ranunculus muricatus, Linn. South Europe.

Isles 2. Miguel, Fayal. Seubert flo. 307. Watson cat. 4. Hunt coll. Probably introduced from Europe, equally with many other such weedy plants.

7. Nigella arvensis, Linn. South Europe.

Isles . . . ? Miguel, &c. "Inter segetes, quibuscum videtur immigrasse;" Seubert flo. 309. "In or near gardens only;" T. C. Hunt msc. "Tout l'archipel;" Drouet cat. no. 8; but not marked as actually seen by himself or companions. (See above, the remarks under *Ranunculus repens*.)

8. Aquilegia vulgaris, Linn. Europe. Britain.

Isle 1. Pico. Seubert flo. 310. Watson cat. 6. Kew list of Godman coll. Drouet cat. 8.

It would seem that all the three or four collectors who have found this plant, observed it on Pico only, where it occurs with white flowers, perhaps with white flowers only.

9. Delphinium Ajacis, "Linn." South Europe.

Isle 1, or more. Flores; Dr. Mackay! "Inter segetes hinc inde;" Seubert flo. 311. "Les champs;" Drouet cat. 9.

Under name of *Delphinium Consolida* in the works of Seubert and Drouet; but *D. Ajacis* (as understood in England) was the species sent to me from Flores by Dr. Mackay, English Vice-Consul there; and in Drouet's Catalogue the plant is not entered as one actually seen, so that his name for it was probably copied from Seubert's Flora.

2. Papaveraceæ.

10. Papaver somniferum, Linn., varr. S.E. Europe? Isles 2. Santa Maria; Hunt coll. Drouet cat. 10. Fayal, a casual escape; Watson coll.

Mr. Hunt sent two forms of this from Santa Maria; neither of them from San Miguel (if I remember rightly), though one is assigned to the latter island in Drouet's Catalogue. One of the forms is glaucous and glabrous, being the same thing as the handsome Poppy formerly often sown as an ornamental annual in our English gardens. The other is less glaucous, with narrower leaves, especially downwards; the peduncles and sepals, as well as the leaves, thinly covered with a bristly pubescence, that of the leaves being chiefly at the midrib, but also terminating the teeth, and in some examples scattered on the under surface; this form likewise being seen occasionally as a half-naturalized weed in our gardens and about waste grounds. The former well corresponds with Dr. Welwitsch's example of "*Papaver somniferum*, L.," in the Flora Lusitanica, Estremadura, 1848. The latter agrees as well with "*Papaver setigerum*, DC.," of Jamin's Plantes d'Algérie, 1851, no. 121. The condition of the plants in Santa Maria is not explained. The glabrous form found by myself in Fayal seemed so little entitled to place among the native and established plants, that it was left unmentioned in my former list. M. Drouet enumerates these two forms as so many true and native species.

11. Papaver Rhœas, Linn. Europe. Britain.
Isles 2. Miguel, Maria. Hunt coll. Drouet cat. 12. Doubtless introduced; and in cultivation as an ornamental plant.

12. Papaver dubium, Linn. Europe. Britain.
Isles 3. Miguel, Fayal, Flores. Watson cat. 7. Hunt coll. Godman coll. Introduced weed?

13. Chelidonium majus, Linn. Europe. Britain.
Isles 2, or more. Miguel, Terceira. Seubert flo. 312. Hunt coll. Godman coll. Drouet cat. 14.

Dr. Seubert writes of this as being found "in cultis et ad vias insularum plurium." Mr. Hunt says of it in San Miguel "in or near gardens only." M. Drouet names Terceira, and adds "et la plupart des autres îles," which seems to be simply a translated repetition of the words from Seubert's Flora, and not any additional confirmation of the alleged fact. The plant was not seen by myself.

3. Fumariaceæ.

14. Fumaria muralis, Sonder. Europe. Britain.
Isles 3. Miguel, Fayal, Flores. Seubert flo. 313. Watson cat. 8. Hunt coll. Kew list of Godman coll.

In Seubert's Flora, equally as in Watson's Catalogue, this is enumerated under the very aggregate Linnean name of "*Fumaria capreolata,*" though as a variety "*minor*" in the Catalogue.

15. Fumaria officinalis, Linn. Europe. Britain.
Isles several? Miguel; Drouet cat. 16. Seubert flo. 314, without named habitat. Kew list of Godman coll.

Dr. Seubert writes of this as occurring "in arvis et inter segetes insularum fere omnium," but without the second number to show it a species collected by Mr. Hochstetter for distribution. M. Drouet marks it as a species seen by himself in San Miguel, while the *F. muralis* is not marked as having been seen in any of the Isles. Mr. Hunt, a resident in San Miguel, sent me thence the *F. muralis* only. Likely as it may be deemed that *F. officinalis* should be a weed in the Isles, I am led to guess that the two specific names here given may perhaps mean one single species (*F. muralis*).

16. Fumaria micrantha, Lagasca. Europe. Britain. Isle 1. Seen in San Miguel; Drouet cat. 17, the only authority, but the species not unlikely to have occurred as an imported weed from Portugal.

4. Cruciferæ.

17. Matthiola annua, Sweet. South Europe? Isles 3, or more. Miguel, Terceira, Fayal. Seubert flo. 315. Watson cat. 0. Hunt coll. Godman coll. Drouet cat. 18. Probably introduced.

Dr. Seubert says of this "in rupibus ad littora insularum fere omnium." M. Drouet follows suit by the corresponding words "tout l'archipel, sur les rochers du littoral." On the cliffs near the town of Horta, in the Isle of Fayal, it appeared to me to be simply a garden escape.

Drouet's Catalogue enumerates also "*Matthiola incana*, R. Br., Spr., var. *affinis*. *M. maderensis*, Lowe.—*Hab.* . . . (Hartung)." I have seen nothing like *M. maderensis* from the Isles; and while adhering here to the nomenclature of Seubert, my suspicion is that the Azore plants are truly forms of *M. incana*; but I possess them in fragments only.

18. NASTURTIUM OFFICINALE, R. Br. Europe. Britain.
Isles 3, or more. Miguel, Fayal, Flores. Seubert flo. 317. Watson cat. 10. Hunt coll. Godman coll. Drouet cat. 21.

Seubert writes of this as being found "ad rivulos omnium insularum," an indication which will not warrant the addition of any other names to the three islands mentioned on other authority.

19. NASTURTIUM FLEXUOSUM, Seubert. Azores only?
Isle 1. Flores; Seubert flo. 316. 61. Drouet cat. 20. This remains unknown to me. Is it any thing else than a state of the variable *N. officinale*?

20. CARDAMINE CALDEIRARUM, Guthnick. Azores only.
Isles 5. Miguel, Pico, Fayal, Flores, Corvo. Seubert flo. 318. 60. Watson cat. 9. Hunt coll. Drouet cat. 23.

Sufficiently distinct from all the British species; coming between *C. amara* and *C. sylvatica* in general appearance; nearest to the latter in technical characters. Unlike most of the other peculiarly Azoric plants, this one bears the climate of England quite well, maintaining itself self-sown and quasi-spontaneously in my own garden in Surrey, more especially in spots where it gets the benefit of occasional waterings in dry weather, as among Ferns.

21. CARDAMINE HIRSUTA, Linn. Europe; Britain.
Isles 2. Miguel; Hunt coll. Terceira; Godman coll. Banksian Herbarium, collected by Masson in San Miguel; Dr. Charles Lemann. Drouet cat. 22.

22. BARBAREA PRÆCOX, R. Brown. Europe?
Isles 2. Miguel, Fayal. Watson cat. 10 *bis*. Hunt coll. Godman coll.

Drouet's Catalogue quotes Watson for this plant in Fayal, as for a species not found by himself in the Isles. But he adds also "25. *Barbarea intermedia*, Bor. (*B. angustana*, Boiss.)—Hab. San Miguel; mai." This latter is given as a second species actually seen. Mr. Hunt's specimens from San Miguel are assigned by myself to *B. præcox*, the so-called "American cress" of our English gardens. The plant labelled as "*Barbarea intermedia*, Bor.," by English botanists is different, and has not been seen by myself either in or from the Isles.

23. Sisymbrium officinale, Scop. Europe. Britain.
Isles 4. Miguel, Maria, Fayal, Flores. Watson cat. 11. Hunt coll. Godman coll. Drouet cat. 26.

24. Sisymbrium Irio, Linn. Europe. Britain.
Isle 1. Fayal; Drouet cat. 27, the sole authority, but the name marked as that of a plant actually seen.

25. Sinapis nigra, Linn. Europe. Britain.
Isles 2, or more. Fayal, Flores. Seubert flo. 322. Watson cat. 12. Godman coll.

Dr. Seubert vaguely says "in incultis, ad muros hinc inde." Not sent by Hunt, nor seen by Drouet. Perhaps only a casual escape from cultivation.

26. Alyssum maritimum, Linn. South Europe.
Isles 2, or several. Miguel; Hunt coll. Terceira; Godman coll. "Secus littora omnium insularum;" Seubert flo. 319. "Tout l'archipel;" Drouet cat. 29.

27. Senebiera pinnatifida, DC. America. (Europe.)
Isles 3, or more. Miguel, Maria, Fayal, &c. Seubert flo. 323. Watson cat. 17. Hunt coll. Godman coll. Drouet cat. 30.

Dr. Seubert writes of this "frequentissime ad vias et margines agrorum insularum omnium," which ought to imply at least the islands of Terceira and Pico, in addition to those named. I did not observe it in Flores or Corvo. The remaining two islands (San Jorge and Graciosa) have been very slightly explored by botanists. The plant is now widely spread over South Europe, and is extending its area in England; but it is supposed by the De Candolles to have been imported into the old Continent from America.

28. Senebiera coronopus, Poir. Europe. Britain.
Isle 1. Terceira (Morelet); Drouet cat. 31, apparently the sole authority. Not an unlikely plant to occur in the Isles, especially as an introduction to them from Europe.

29. Lepidium virginicum, Linn. America.
Isles 3. Terceira, Pico, Fayal. Seubert flo. 321. Watson cat. 16. Hunt coll. Godman coll. Drouet cat. 32.

As indicated by the specific name, this should be an American species originally, one nearly allied to the European *L. ruderale*. It is fully established in Fayal, where it was seen by myself, and subsequently by Mr. Hunt. Not yet reported from the chief island (San Miguel).

30. Capsella Bursa-Pastoris, Mœnch. Eur. Britain.
Isles 3. Miguel; Hunt coll. Terceira; Godman coll. Flores; Dr. Mackay! Drouet cat. 33. Perhaps neglected rather than non-observed by the other collectors.

31. Rapistrum perenne, Berg. South Europe.
Isles 1, or 2. Miguel? Fayal; Watson cat. 13.
It is doubtful whether any of the specimens of *Rapis-*

trum which were collected in San Miguel by Mr. Hunt are properly to be assigned to this species, although some of them may have been so labelled. Following the distinctive characters given in the Prodromus of De Candolle, those from Fayal may be rightly thus named; but I do not clearly understand the *orientale* and *Linneanum,* apart from the two species here reported.

32. Rapistrum rugosum, Berg. South Europe.

Isles 2. San Miguel; Hunt coll. Graciosa; Drouet cat. 35.

Drouet's Catalogue adds also a third species, under the name of "*Rapistrum orientale,* DC." Perhaps that may be a smooth-fruited form of the present species. In my herbarium are two specimens from Madeira, labelled by the late Dr. Charles Lemann respectively *R. perenne* and *R. rugosum,* both of which (in the absence of fruit-pods on the former) appear to my eyes exactly the same thing with the Azore plant, which here I name as *R. rugosum.*

33. Cakile americana, Nuttall. America.

Isle 1. Fayal; Watson cat. 15, 1842; also T. C. Hunt coll. 1847. Not found elsewhere, nor by other collectors.

This appears under the name of *Cakile maritima* in the lists of Azore plants. I noted it as a variety distinguished by the ovate and emarginate upper joint of the silicle, the lower joint usually seed-bearing also. Professor C. C. Babington suggested to me that it is really *C. americana,* whether properly to be held a distinct species or otherwise. In the Prodromus of De Candolle, the diagnosis is made to depend on the upper joint of the pod being "ensiform" in *maritima,* "ovate-acute" in *americana.* In the Fayal examples, and in others raised from their seed in England, the upper joint is decidedly emarginate, and otherwise less acute than in the European *C. maritima.* My only herbarium example from America has the upper joint of the still immature silicle produced into a point, blunt, but scarcely emarginate. So far the plant of Fayal is really more distinguishable from European examples than is this one American (Boston?) example. The leaves of the Azorean and American examples are nearly alike, and much less sinuate than those of ordinary *C. maritima.* On distant recollection the plants in Fayal were more erect and shrub-like than those of the English coast. The Azore plant has hitherto been found only in one spot in one isle—namely, the little sandy bay or nook at Porto Pym, near the town of Horta, in Fayal. Phytogeographers and Darwinists can apply this fact as to them may seem meet.

34. Raphanus raphanistrum, Linn. Europe. Britain. Isles 2, or 3. Miguel, Maria; Drouet cat. 38. Fayal, uncertain; see the next species.

35. Raphanus landra, Moretti. South Europe? Isles 2. Fayal, Flores. Watson cat. 14 (in part), under the name of *Raphanistrum.*

Probably there was a confusion between this present and the preceding species among the few specimens distributed in 1842–43, under name of *Raphanistrum,* some being rightly so labelled, others belonging to the present species. In reference to the uncertainty, M. Drouet states that his examples from San Miguel and Santa Maria evidently belong to *Raphanistrum*; as evidently, I think, the specimens raised in England from Azore seed must be referred to *Landra,* as figured by Delessert (Table 94), and likewise as distributed by L. Kralik in specimens from Sicily, with the date of 1847 on their labels. But according to

Dr. Grenier, pronouncing on garden examples of my Azore plant, it is *Enarthrocarpus strangulatus*, originally from Egypt, an idea with which I cannot concur.

5. Resedaceæ.

36. Reseda luteola, Linn., var. crispata.

Europe. Britain.

Isles 7. Miguel, Maria, Terceira, Graciosa, Pico, Fayal, Flores. Watson cat. 18. Hunt coll. Godman coll. Drouet cat. 39.

It is remarkable that the present plant should have been omitted from the Flora by Seubert, seeing that it has been found in so many of the islands. This omission, in connexion with various internal evidences in the Flora, leads to a supposition that Messrs. Hochstetter and Guthnick sometimes disregarded plants supposed to be ordinary European species. The variety is South European, the species extending northward into Scandinavia.

37. Reseda "macrosperma, Reich." South Europe.

Isle 1. San Miguel. "In vineis;" Seubert flo. 324. 43. Hunt coll. Not observed by other collectors. If this is rightly named, the earlier specific name of the plant would seem to be "*media*," according to Dr. Mueller, in De Candolle's Prodromus.

6. Violaceæ.

38. Viola palustris, Linn. Europe; Britain.

Isles 2. Pico, Flores. Watson cat. 20. Apparently not found by any other collector, unless mistaken by them for *odorata*.

Being absent from Madeira and Canaries, this boreal

Violet may be held one of the special affinities between the flora of Europe and that of the Isles.

39. VIOLA ODORATA, Linn. Europe. Britain.

Isles 2. Fayal, Flores. Seubert flo. 327. 17. Watson coll. 19.

Introduced?

Seubert describes the habitat of this plant, "in paludosis insulæ Flores, mense Junio fructibus maturis lecta," which suggests the idea of some error between the present and the preceding species. On the contrary, Drouet describes it as being "common in the woods of the mountains," but without prefixing the mark to indicate it a plant actually seen by himself or companion. What entitles him to locate it in the mountain woods I am unable to say. Mr. Webb held it to be the *Viola maderensis* of Lowe, "but it is only one of the hundred forms of *V. odorata.*"

40. VIOLA TRICOLOR, Linn. Europe. Britain.

Isle 1. Flores; Dr. Mackay; probably an escape from gardens there, although ornamental gardening was in a very rudimentary state in Flores in 1842.

M. Drouet erroneously reports this from "San Miguel (Watson)," my herbarium example having been sent from Flores by Dr. Mackay, as expressly stated in the 'London Journal of Botany,' vol. vi. p. 381.

7. POLYGALACEÆ.

41. POLYGALA VULGARIS, Linn. Europe. Britain.

Isle 1. Pico. Seubert flo. 344. 67. Watson coll. Drouet cat. 77.

M. Drouet records also *Polygala depressa*, no 78, as found by M. Arthur Morelet in Terceira. Only one fragment from Pico is in my herbarium, which may be the *depressa*, although I cannot certainly say that it is such; indeed I am often at a loss how to divide more complete British specimens between *depressa* and *vulgaris* (limited).

8. Frankeniaceæ.

42. Frankenia pulverulenta, Linn. South Europe. Isles 5. Miguel, Terceira, Pico, Fayal, Flores. Seubert flo. 326. 26. Watson cat. 23. Hunt coll. Godman coll. Drouet cat. 44.

43. Frankenia ericifolia, "Chr. Smith." Canaries. Isle 1. Corvo. Seubert flo. 327. 25. Watson cat. 22. Godman coll.

This is clearly the same with the Canary plant, as represented by Bourgeau's specimens, no 5. If the species had been recognized as a plant of Madeira by Lowe's Manual Flora, I should have been led to refer to it a very poor specimen from Madeira, labelled by Dr. Lemann as *Frankenia cæspitosa*, Lowe; which will otherwise go to the polymorphic and inappropriately named *F. lævis* as the alternative.

9. Caryophyllaceæ.

44. Silene inflata, Smith?, var. Europe. Britain. Isles 3. Miguel, Pico, Flores. Seubert flo. 335. 86. Watson cat. 25 (*S. maritima*). Hunt coll. Godman coll. Drouet cat. 59 (*S. inflata*, variety *rupicola*, Bor.).

Possibly both *inflata* and *maritima* occur in the Isles. The examples sent by Mr. Hunt from San Miguel since my former catalogue was published cannot be referred to *S. maritima*, nor do they quite correspond with our British *S. inflata*; possibly distinct from both.

45. Silene gallica, Linn. Europe. Britain.

Isles 4, or more. Miguel, Maria, Fayal, Flores. Seubert flo. 336. 35. Watson cat. 24. Hunt coll. Drouet cat. 60. Kew list of Godman coll.

In the Flora Azorica this is named *lusitanica*. In Drouet's Catalogue both names are enumerated as those of two species, both of them seen by himself,—*lusitanica* in Santa Maria, *gallica* in San Miguel. I confess my inability to distinguish satisfactorily from each other the *lusitanica, gallica,* and *anglica,* and thus adopt the middle name as applicable to all three. The *Silene Armeria* is an introduced species, casually subspontaneous.

46. Cerastium azoricum, Hochst. Azores only.

Isles 2. Flores, Corvo. Seubert flo. 333. 88. Watson cat. 26. Drouet cat. 57.

This is a pretty species, with more conspicuous flowers than would be supposed from the figure of it in Flora Azorica, which has doubtless been taken from a dried and much shrivelled specimen. Almost equally hardy with *Cardamine caldeirarum* in respect of bearing cold. But seeding less freely, and ill-fitted to bear dryness, it soon became extinct in my garden in Surrey.

47. Cerastium glomeratum, Thuill. Europe. Britain.

Isles 5. Miguel, Maria, Terceira, Fayal, Flores. Seubert flo. 334? Watson cat. 27. Hunt coll. Godman coll. Drouet cat. 55 or 56?

Under name of *C. vulgatum* in the Flora Azorica, with the indication "in agris hortis et incultis," which seems to describe the usual places of growth of *C. glomeratum* in Europe. Among English botanists these two names are held synonymous; usually not so among the continental botanists, several of whom make *viscosum* synonymous with *glomeratum*.

48. Cerastium triviale, Link. Europe. Britain. Isles 2, or 3. Miguel, Fayal, Flores. Watson cat. 28. Hunt coll. Drouet cat. 56 or 55?

The names *vulgatum* and *viscosum* have been so much crossed or alternately applied between this and the preceding species, that it has become almost impossible to say what is really intended by either in a list of mere names; hence the uncertainty respecting the nos. in Drouet's Catalogue, where those two names are adhered to.

49. Stellaria media, Withering. Europe. Britain. Isles 3, or more: Miguel, Fayal, Flores. Watson cat. 29. Hunt coll. Godman coll. Probably disregarded by the other collectors and authors.

50. Mœhringia muscosa, Linn. Middle Europe. Isle 1. Terceira (Morelet); Drouet flo. 54. Not reported by the other collectors. Perhaps an error?

51. Sagina procumbens, Linn. Europe. Britain. Isles 4. Miguel, Maria, Fayal, Flores. Seubert flo. 331. Watson cat. 32. Hunt coll. Godman coll. Drouet cat. 49. "Everywhere . . . . in damp places;" Seubert.

It is probable that some examples of *Sagina apetala* were mingled with others of the present species in my own collection, and that both species will be found to occur in the Isles.

52. Spergula arvensis, Linn. Europe. Britain. Isles 2. San Miguel; Hunt coll. and Drouet cat. 56. Fayal; Godman coll.

53. Spergularia rubra, Presl. Europe. Britain. Isles 5. Miguel, Maria, Terceira, Fayal, Flores. Watson cat. 30. Hunt coll. Godman coll. Drouet cat. 51.

54. Spergularia marina, Auct. Europe. Britain. Isles 2, or more. Miguel, Pico, &c. "*Alsine marina* . . . . frequens in rupibus littoralibus omnium insularum;" Seubert flo. 332 *a*. "*Arenaria marina* . . . . San Miguel, Pico;" Drouet cat. 53.

It would seem likely enough that one or more of the various segregates which have been included under the name of *Arenaria marina*, or of some of its synonyms, may occur in the Isles, though I do not understand what is intended by Seubert and Drouet, apart from the next very dubious species. The name occurs in Dr. Lemann's list for Madeira; but Mr. Lowe does not admit any "*marina*" in the Manual Flora.

55. Spergularia "macrorhiza" ?? South Europe? Isles 4, or more. Miguel, Maria, Fayal, Flores. Seubert flo. 332 (in part), 37. Watson cat. 31. Hunt coll. Godman coll. Drouet cat. 52.

A most puzzling plant. In Seubert's Flora this was mentioned under the name of "*Arenaria macrorhiza*, Req.," and placed as a remarkable variety of *Arenaria* (or *Alsine*) *marina*. Both names are enumerated in Drouet's Catalogue as those of two species additional to *Arenaria rubra*, which last was not mentioned by Seubert. In my own former catalogue, the specific name "*macrorhiza*, Req.," was simply quoted from the Flora Azorica, as its application to the specimens appeared to my own judgment unsatisfactory, while I was not then prepared to deny it. I now suppose the Mediterranean *macrorhiza* to be the same with the *rupicola* or *rupestris* of English botanists. If so, it is not the same with the Azore plant under consideration. I have seen nothing from any other habitat which I can declare an identical species with that of the Isles. This plant so far diverges from the forms usually assigned to *rubra* or *marina,* in their aggregate applications, that at first I labelled it *Arenaria* (*Rhodalsine*) *procumbens.* From the *Rhodalsine,* however, it is well distinguished by the presence of membranous stipules, which are shorter and wider than usual with those of *rubra* and *marina.* In size of flowers it comes nearer to *rubra*; by its thick perennial root it resembles *macrorhiza* or *rupicola.* From both it differs in its leaves, shorter, wider, less acute, often quite obtuse, and in being covered with a dense viscid pubescence. Unwillingness to encumber this genus with another specific name for a possible variety of *rubra* prevents the description of the Isles' plant here in the character of a new species.

56. Polycarpon tetraphyllum, L. Europe. England. Isles 3, or more. Miguel, Fayal, Flores. Seubert flo. 330. Watson cat. 90. Hunt coll. Drouet cat. 48. All the islands; Seubert.

10. Portulacaceæ.

57. Portulaca oleracea, Linn. Europe. Isles 4. Miguel, Pico, Fayal, Flores. Seubert flo. 328. Watson cat. 89. Hunt coll. Godman coll. Drouet cat. 46.

11. Elatinaceæ.

58. Elatine hexandra, De Cand. Europe. Britain. Isle 1. Flores. Watson cat. 32. Apparently not

found by any other collector in the Isles; being a very small aquatic plant, it is thus likely to escape notice.

12. Hypericaceæ.

59. Hypericum foliosum, Ait. Azores only?

Isles 5, or more: Miguel, Graciosa, Pico, Fayal, Flores. Seubert flo. 343. 87. Watson cat. 38. Hunt coll. Godman coll. Drouet cat. 75 & 76.

This conspicuous and plentiful shrub in the Isles has been variously named (see London Journal of Botany, vol. iii. pp. 587, 588). By Seubert and Drouet the name *Androsæmum Webbianum* (Spach) is applied to it; *Hypericum grandifolium* (Chois.) being added as a synonym. Each author repeats it also, as if a second species, under the name here adopted for it; Seubert simply quoting from Masson; but Drouet boldly giving both names as those of two species actually seen. It does not clearly appear, however, that more than one shrubby species of *Hypericum* exists in the Isles; although the dried specimens from Mr. Hunt showed considerable diversity in foliage. Mr. Lowe denies the specific identity of *foliosum* and *grandifolium* in his Manual Flora of Madeira.

60. Hypericum perforatum, Linn. Europe. Britain.

Isles 1, or 2. Terceira, Fayal? Seubert flo. 340 (in part?), 87. Terceira; Godman coll. Fayal; Drouet cat. 70.

Neither sent by Mr. Hunt from San Miguel nor seen by myself in any of the four islands visited in 1842; but the specimen brought from Terceira by Mr. Godman is certainly this species. Dr. Seubert omits the *Hypericum bæticum*; and by M. Drouet it is only given as a species not seen by himself or fellow travellers. I am thus led to suppose that the so-called *perforatum* of the two authors named was really the *bæticum,* in whole or in part. This seems all the more likely because Dr. Seubert records his "*perforatum*" as found "ubique in lapidosis collinis,"—a suspicious indication for a showy species not found in San Miguel by Mr. Hunt, nor seen in any of the four islands visited by myself, while *H. bæticum* has been found in three of the latter and also in San Miguel.

61. Hypericum bæticum, Boiss. S. Europe. England. Isles 4. Miguel; Hunt coll. Pico; Baron do Castello de Paiva in Kew Herbarium. Fayal, Flores; Watson cat. 39.

This is the *Hypericum decipiens* of my former catalogue in the London Journal of Botany, vol. iii. pp. 588, 589. Probably it is the *Hypericum undulatum* of Schousboe, although Reichenbach's figure of that plant would be a very faulty representation of the Azore species. The alleged *H. tetrapterum* of Fries, enumerated by Drouet as having been actually seen in Flores, is almost certainly the present species, which is recorded as one not seen by himself in the Isles. See the remarks on the preceding species. It is still confused with *H. quadrangulum* in the Manual Flora of Medeira.

62. Hypericum humifusum, Linn. Europe. Britain. Isles 6. Miguel, Maria, Terceira, Pico, Fayal, Flores. Seubert flo. 341. Watson cat. 40. Hunt coll. Godman coll. Drouet cat. 72.

63. Hypericum elodes, Linn. West Europe. Britain. Isle 1. San Miguel; sent thence by Mr. Hunt in 1847; but apparently not found by any non-resident collector.

This is geographically interesting as one of the plants which specially connect the Azoric and West-European floras, without extending its area into the other Atlantic Isles; nor is it included in Munby's plants of Algeria.

13. Malvaceæ.

64. Malva parviflora, Linn. South Europe.
Isles 2, or more. Miguel; Hunt coll. Fayal; Watson cat. 36. Not found by the other collectors, or mistaken by them for *M. rotundifolia*.

65. Malva nicæensis, All. South Europe.
Isle 1, or more. Fayal; Watson cat. 35, under the name of *Malva rotundifolia* erroneously. It was found in Fayal by Mr. Godman also, and possibly it has been placed in the Kew Herbarium from him under the same erroneous name. See the next species.

66. Malva "rotundifolia, Linn." Europe. Britain.
Isle 1, or more by report. "In incultis plurium insularum;" Seubert flo. 338. Fayal; Watson cat. 35. Kew list of Godman coll. Drouet cat. 66.

Here we seem to have ample authority for admitting this species among Azoric plants; and yet I now much doubt it fairly belonging to the Isles. My own herbarium specimen thus named from Fayal is really *Malva nicæensis*; and the same is the case with one given to me by Mr. Godman, so far as can be declared in the absence of fruit on it. On the other hand, the Flora Azorica makes no mention of *M. parviflora*, while Drouet enumerates it only with my own name added, instead of his mark used to indicate plants seen by himself or companion. The inference thus fairly seems to be that *nicæensis* and *parviflora*, one or both, have hitherto been mistaken for *rotundifolia*; this latter plant properly so named not occurring in the Isles.

67. Lavatera sylvestris, Brot. South Europe.
Isles 6. Miguel, Terceira, Graciosa, Pico, Fayal, Flores. Seubert flo. 337. 63. Watson cat. 34. Hunt coll. Godman coll. Drouet cat. 65.

68. Sida rhombifolia, Linn. Madeira. Canaries.
Isles 4. Miguel, Pico, Fayal, Flores. Seubert flo. 339. 65. Watson cat. 37. Hunt coll. Godman coll. Drouet cat. 68.

Perhaps introduced; as it may have been also to Madeira and the Canaries.

14. Geraniaceæ.

69. Erodium malacoides, Linn. South Europe.
Isles 2. Miguel, Pico. Seubert flo. 348. 66. Watson cat. 43. Hunt coll. Drouet cat. 88.

70. Erodium moschatum, Willd. Europe. England.
Isles 2. Miguel; Hunt coll. Terceira; Drouet cat. 88; but not marked as a plant seen by himself or companion.

71. Geranium rotundifolium, L. Europe. England.
Isle 1. San Miguel; Drouet cat. 86. Kew list of Godman coll.

72. Geranium molle, Linn. Europe. Britain.
Isles 3. Miguel; Hunt coll. Fayal; Watson cat. Terceira; Godman coll. Drouet cat. 85.

73. Geranium dissectum, Linn. Europe. Britain. Isle 1, or more. Miguel, &c. Several islands; Seubert flo. 349. Miguel; Hunt coll. and Godman coll. All the islands; Drouet cat. 84.

74. Geranium Robertianum, Linn. Europe. Britain. Isles 4, or more. Miguel, Terceira, Fayal, Flores. Seubert flo. 350. Watson cat. 41. Hunt coll. Godman coll. "Hab. tout l'archipel;" Drouet cat. 87.

15. Oxalidaceæ.

75. Oxalis corniculata, Linn. Europe. (England.) Isles 4, or more. Miguel, Maria, Fayal, Flores. Seubert flo. 351. Watson cat. 44. Hunt coll. Godman coll. Drouet cat. 90. "Hab. ad vias et in ruderatis insularum fere omnium;" Seubert.

Oxalis purpurea was sent from San Miguel by Mr. Hunt; and Mr. Godman found the same or some other purple-flowered species in Fayal: doubtless escaped from gardens originally in both islands.

16. Rutaceæ.

76. Ruta bracteosa, De Cand. South Europe. Isles 3. Miguel, Maria, Terceira. Seubert flo. 347. Hunt coll. Godman coll. Drouet cat. 83. Native?

17. Ilicaceæ.

77. Ilex Perado, Aiton. Madeira. Canaries? Isles 5. Miguel, Terceira, Pico, Fayal, Flores. "All the islands;" Seubert flo. 345. 7. Watson cat. 45. Hunt coll. Godman coll. Drouet cat. 79. This is not the

Ilex canariensis; but it may be a smaller state of Webb's *Ilex macrophylla*.

18. Rhamnaceæ.

78. Rhamnus latifolius, Hérit. Madeira? Isles 4, or more. Miguel, Maria, Fayal, Flores. Seubert flo. 346. 16. Watson cat. 46. Hunt coll. Drouet cat. 80. "All the islands;" Seubert.

"*Rhamnus pubescens*, Banks, MS.—Hab. Terceira, dans les bois des Garridas (Morelet). Rare;" Drouet flo. 81. What is this?

19. Anacardiaceæ.

79. Rhus coriaria, Linn. South Europe. Isles 2. Pico, Flores. Watson cat. 47. Not reported by the other collectors; and though seen in some quantity on a low hill near the coast of Pico, it might have been planted there originally for economical purposes, since it is said (Lowe's Manual Flora) to be used in Madeira for tanning.

20. Leguminiferæ.

80. Sarothamnus scoparius, Wimm. Europe. Britain. Isles 5. Miguel, Terceira, Fayal, Flores, Corvo. Seubert flo. 365. 68. Watson cat. 49. Drouet cat. 116, "tout l'archipel."

It may be questioned whether this shrub is truly indigenous in the Isles. The *Spartium junceum* is more evidently an alien there, though often cultivated, and half-wild in various spots. Drouet enumerates also the *Ulex europæus* as seen in San Miguel and Santa Maria, and the *Ulex nanus* on the mountains in Flores; both these likely being introduced shrubs.

81. ONONIS ARVENSIS ("Linn."), Auct. Angl.
Europe. Britain.

Isle 1. San Miguel; Hunt coll. 1844. If more examples of this were sent at the date, I have retained from them only a single branch for my herbarium, which is entirely without spines, and corresponds well with Danish and German specimens labelled "*Ononis procurrens*, Wallr.," by Dr. Petit and Herr Hornung.

82. TRIGONELLA ORNITHOPODIOIDES, De Cand.
Europe. Britain.

Isle 1. Sent from San Miguel by T. C. Hunt in 1847. Apparently not found by any other collector; but it is a small plant, easily overlooked.

A South-European plant; absent from Madeira, Canaries, Algeria; extending northward to Denmark and Britain, on the west side of Europe.

83. MEDICAGO LUPULINA, Linn. Europe. Britain.

Isles . . . . ? "Ad vias, in arvis agrisque insularum fere omnium;" Seubert flo. 366. "Tout l'archipel;" Drouet cat. 120.

84. MEDICAGO LAPPACEA, "Desr. in Lam." S. Europe.

Isles 3. Miguel; Hunt coll. Maria; Drouet cat. 121. Terceira; Seubert flo. 367 (*lappacea*) and 368 (*pentacycla*).

The San Miguel specimens quite correspond with those of *M. lappacea* from Madeira &c. See Lowe's Manual Flora of Madeira for serviceable remarks on this species.

85. MEDICAGO DENTICULATA, Willd. Europe. England.

Isles 1 or 2. Fayal; Watson cat. 72. Godman coll. Terceira also?

My own examples collected in Fayal, and plants raised from seeds brought thence to England, are the species usually intended by French and English botanists under the name here used. The pods vary in number from 3 to 10 on a peduncle, thus well bearing out the specific name of "*polycarpa*" taken up by the Authors of the Flore de France to cover the *denticulata, apiculata,* and *tuberculata,* united together as a single species.

86. MELILOTUS PARVIFLORA, Desf. South Europe. Isles 3. Miguel, Terceira, Corvo. Seubert flo. 369. 75. Watson cat. 59. Hunt coll.

87. TRIFOLIUM ANGUSTIFOLIUM, Linn. South Europe. Isles 2 or 3. Maria?, Terceira, Fayal. Seubert flo. 369. 75. Watson cat. 60. Hunt coll. Godman coll. Santa Maria; Drouet cat. 125; but not marked as found by himself or fellow travellers.

88. TRIFOLIUM ARVENSE, Linn. Europe. Britain. Isles 5. Miguel, Terceira, Graciosa, Fayal, Flores. Watson cat. 61. Hunt coll. Godman coll. Drouet cat. 134.

89. TRIFOLIUM LIGUSTICUM, Linn. South Europe. Isles 4, or more. Miguel, Maria, Fayal, Flores. Seubert flo. 370. 73. Watson cat. 62. Hunt coll. Godman coll. Drouet cat. 126. Hab. "fere ubique;" Seubert flora.

90. TRIFOLIUM SCABRUM, Linn. Europe. Britain. Isles 3: Miguel, Terceira, Fayal. Seubert flo. 373. 74. Watson cat. 66. Hunt coll. Godman coll. Drouet cat. 129.

91. Trifolium striatum, Linn. Europe. Britain. Isles 1, or 2. San Miguel; Hunt coll. Terceira; Drouet cat. 141; but not marked as a species actually seen by himself or companion. What, then, is the real authority for the habitat of Terceira?

Trifolium rariflorum of Welwitsch appears to be a variety of *T. striatum*; it was sent from San Miguel.

92. Trifolium maritimum, Huds. Europe. England. Isles 1, or 2. Santa Maria; Hunt coll. Terceira; Drouet cat. 138; not marked as seen by himself or companion.

93. Trifolium lappaceum, Linn. South Europe. Isle 1. Fayal; Seubert flo. 372. 72. Included in the Kew list of Mr. Godman's collection; but no specimen of it came under my own observation there.

94. Trifolium repens, Linn. Europe. Britain. Isles 3, or more. Miguel, Fayal, Flores. Everywhere; Seubert flo. 276. Watson cat. 63. Hunt coll. Kew list of Godman coll. Drouet cat. 132.

95. Trifolium glomeratum, Linn. Europe. England. Isles 3. Miguel, Fayal, Flores. Seubert flo. 374. 71. Watson cat. 64. Hunt coll. Godman coll. Drouet cat. 131.

96. Trifolium suffocatum, Linn. Europe. England. Isle 1. Pico. Seubert flo. 374. Watson cat. 65. See the next species.

97. TRIFOLIUM CERNUUM, "Brotero." Portugal.

Isles 2. San Miguel, 1849; T. C. Hunt coll. Pico, 1842; Watson coll.

The name of this plant is taken from the label of no. 337 of Dr. Welwitsch's Flora Lusitanica. The examples from San Miguel and Pico appear to be quite the same. One picked by myself on Pico had been at first overlooked among others of *T. suffocatum*.

98. TRIFOLIUM RESUPINATUM, Linn. South Europe.

Isle 1. Santa Maria; Hunt coll. Godman coll. Miguel; Drouet cat. 139, perhaps an error of place.

99. TRIFOLIUM SUBTERRANEUM, L. Europe. England.

Isles 2. Miguel; Hunt coll. Graciosa (Hartung); Drouet cat. 127.

100. TRIFOLIUM PROCUMBENS, L. Europe. Britain.

Isles 3. Miguel, Pico, Fayal. Seubert flo. 377. Watson cat. 67. Hunt coll. Godman coll. Drouet cat. 135 (*T. campestre*).

101. TRIFOLIUM MINUS, Relhan. Europe. Britain.

Isles 3, or 4. Miguel, Fayal, Flores. Watson cat. 68, under name of *Trifolium filiforme*. Hunt coll. Maria; Drouet cat. 137; but no authority cited, and not marked as having been actually seen by himself or fellow travellers.

This is the plant usually labelled "*Trifolium filiforme*, Linn.," by the botanists of the Continent; but it is not the *T. filiforme* of English botanists, unless by mistake so labelled.

102. Lotus major, Scop. Europe. Britain. Isles 4. Miguel, Terceira, Fayal, Flores? Seubert flo. 382 *a*. Watson cat. 69. Hunt coll. Godman coll. Drouet cat. 147 & 149.

In the Catalogue by Drouet two species are enumerated,—147. *Lotus uliginosus,* Schk., inhabiting Terceira (Morelet) and Miguel (Hartung), and 149. *Lotus major,* Scop., seen by the Author in Fayal and Flores; the *Lotus creticus* being interposed between them as no. 148. Still, it is presumed, this must only show that he mistakes two synonyms as meaning two different species.

103. Lotus corniculatus, Linn. Europe. Britain. Isles 3. Occasionally; Seubert flo. 382. Maria, Terceira, Pico; Drouet cat. 146. Perhaps my own herbarium examples from Flores might be assigned to this rather than to the preceding species.

104. Lotus angustissimus, Linn. Europe. England. Isles 4. Miguel, Graciosa, Fayal, Flores. Seubert flo. 380. 82. Watson cat. 70. Hunt coll. Kew list of Godman coll. Drouet cat. 144.

105. Lotus hispidus, Desf. Europe. England. Isles 5. Miguel, Maria, Terceira, Fayal, Flores. Seubert flo. 379. 88. Watson cat. 70 *bis*. Godman coll. Drouet cat. 145.

106. Lotus parviflorus, Desf. South Europe. Isles 3. Miguel; Hunt coll. Fayal; Watson cat. 71, and Hartung in Drouet cat. 143. Flores; Dr. Mackay! Kew list of Godman coll.

Dr. Seubert doubts this plant being rightly named, alleging that Guthnick's examples belonged to *L. hispidus*. Although much alike, I regard the two species as really distinct, and I certainly collected both.

107. Lotus creticus, Linn. South Europe.
Isle 1. Terceira. Seubert flo. 381. 81. Kew list of Godman coll. Drouet cat. 148.

108. Pedrosia macrantha, Lowe? Madeira.
Isle 1. Santa Maria; Hunt coll. Apparently overlooked by other collectors there, although a remarkable plant.

Whether this is the *macrantha* or *argentea* of Lowe, I cannot absolutely determine, though the long and decidedly stalked legumes seem to place the Azore specimens under the former name.

109. Arthrolobium ebracteatum, DC. Eur. Engl.
Isles 5, or more. Miguel, Maria, Terceira, Fayal, Flores. Seubert flo. 390, 84. Watson cat. 74. Hunt coll. Godman coll. Drouet cat. 165.

110. Ornithopus perpusillus, L. Europe. Britain.
Isles 4. Miguel, Maria; Hunt coll. Flores, Corvo; Watson cat. 73. Maria, Flores; Drouet cat. 466. Godman coll.

111. Ornithopus roseus, Dufour. South Europe.
Isle 1. Terceira; Seubert flo. 391. The correctness of this habitat is confirmed by specimens collected there by Mr. Godman in 1865.

It is curious that the present apparently quite local species should have been the only one mentioned in the Flora Azorica; the more widely diffused *O. perpusillus* having been overlooked. *Ornithopus sativus* of Brotero apparently included two or more species under the single name.

112. Ornithopus compressus, Linn. South Europe. Isle 1. Santa Maria; T. C. Hunt coll. Not from San Miguel, as it is erroneously located in Drouet's Catalogue.

113. Ervum (Vicia) hirsutum, L. Europe. Britain. Isles 3. Miguel, Fayal, Flores. Watson cat. 58. Hunt coll. Godman coll.

114. Ervum (Vicia) gracile, Loisel. Europe. England. Isles 2. Miguel, Fayal. Seubert flo. 384. 86 *b*. Watson cat. 57. Hunt coll. Drouet cat. 152.

Ervum Lens and *Ervum monanthos*, enumerated in the Catalogues, are probably casual escapes from cultivation; and possibly these two names intend the same thing.

115. Vicia Dennesiana, Watson, MS. Azores only. Isle 1. San Miguel; T. C. Hunt coll. (Misprinted "*Durneriana*" in Drouet's Catalogue.)

This is a remarkable plant in itself, and in its history so far as hitherto known; for chance only appears to have saved it from becoming an extinct species almost immediately after it became known at all. It was found by Mr. Hunt "on the mountains at the east end of the island, growing on damp earthy precipices;" but in one spot only, from which it has since disappeared through a landslip. At the time of writing this page, the letter from Mr. Hunt, which conveyed a more detailed account of the discovery and disappearance of the *Vicia* has been itself unfortunately lost or mislaid. Mr. Hunt unsuccessfully sought for the plant elsewhere in the same neighbourhood; and no other collector has found it in any of the Isles. Some years ago I searched the rich herbarium of Sir William Hooker without seeing any plant with which this one could be identified, or even very closely approximated. It has flowered with me occasionally; and cultivated specimens have been distributed to botanists, labelled with the manuscript name of *Dennesiana,* adapted from the name of Mr. G. E. Dennes, who was Honorary Secretary to the Botanical Society of London, at the time when some native specimens were sent by Mr. Hunt for distribution through that Exchange Club. The garden treatment had been to raise the plants from seeds sown in flower-pots, to give them the protection of a frame sheltered from frost in winter, and to put out the grown plants into the open garden ground in the following summer, to flower and seed, a succession being kept up by resowings. The severe frost of May 22 and 24, 1867, in Surrey, proved almost fatal to the small stock of plants then on hand. The roots were repotted; and one of the three is flowering weakly in the present (too dry) summer of 1868, and may perhaps be figured and described in the Botanical Magazine. Meantime the subjoined description will afford a diagnosis of the species. It is made chiefly from the specimens of the plant placed in my own herbarium, where its nearest analogue in general appearance is *Vicia villosa,* although it is distinct at first glance from that and every other *Vicia* known to me.

VICIA DENNESIANA (Watson in sched.). Perennis scandens sericeo-pilosa, caule sulcato subquadrangulari, foliis alternis sessilibus, foliolis numerosis (16–24) oblongis mucronulatis alternis suboppositisve subtus sericeo-pubescentibus supra glabris reticulato-venosis, cirrhis ramosis, stipulis semisagittatis dentatis, racemis multifloris, folium subæquantibus brevioribusve, floribus magnis numerosis (10–20) laxiusculis colore mirabiliter variantibus, dentibus calycis tubo brevioribus, inferiore angusto superiores breves latos paulo excedente, vexillo abbreviato alis patulis subreflexis breviore, leguminibus glabris compressis lineari-oblongis, longitudine latitudinem quater–quinquies excedente.

The changes in the colour of the flowers are remarkable. In the early bud they are of a fine purple; and being an inch long, and numerous in the racemes, they lead to the expectation of a handsome appearance. But as they expand, or even earlier, the colour fades to a dull slate, and finally to a dingy fawn, which becomes a sort of russet in drying for the herbarium. The standard is small and curiously compensated by the spreading wings which curve outward and backward, and thus appear like a divided standard.

116. Vicia albicans, Lowe. South Europe?

Isles 2. Miguel, Fayal. Seubert flo. 386. Watson cat. 56. Hunt coll. Drouet cat. 164.

Almost certainly the *Vicia atropurpurea* (Desf.) of South Europe and North Africa. The name of *Vicia albicans* (Lowe) is repeated for uniformity with the Flora and two Catalogues cited, and because Mr. Lowe (Manual Flora of Madeira) still declares his species distinct from true *atropurpurea* of the Canaries. I do not myself see any characters by which the Azore specimens can be separated from the plants which are held to be truly *Vicia atropur-*

purea; for instance, Kralik's plants of Corsica, no. 569; Bourgeau's plants of Spain, no. 181; Bourgeau's plants of Spain and Portugal, no. 1856. My solitary specimen of the Madeira plant has the raceme 4- or 5-flowered, and it well corresponds with those from the Isles. The wild Azore examples were equally silky-pubescent with that of Madeira; but garden examples, raised in Surrey from Azore seeds, were larger and much less pubescent than the wild plants; in each the racemes bearing 4 or 5 flowers, rarely 2 or 3 only.

117. VICIA ANGUSTIFOLIA, Roth. Europe. Britain. Isles 2, or 3. Miguel, Fayal. Seubert flo. 385. 86 *a*. Watson cat. 55. Hunt coll. Kew list of Godman coll. *V. sativa* in Santa Maria, by Drouet cat. 158.

Some of the San Miguel specimens would be assigned to *V. angustifolia*, if seen apart; while others might possibly have been labelled as *V. sativa* under like circumstances.

118. VICIA BITHYNICA, Linn. Europe. England. Isle 1. San Miguel; T. C. Hunt coll. Probably local, as not found by the other collectors.

119. LATHYRUS APHACA, Linn. Europe. England. Isles 4. Miguel, Pico, Fayal, Flores. Seubert flo. 387. Watson cat. 50. Hunt coll. Kew list of Godman coll. Drouet cat. 155.

120. LATHYRUS SATIVUS, Linn. South Europe. Isle 1. Fayal. Seubert flo. 388. Watson cat. 52. Probably an introduced plant.

Lathyrus annuus is named in the Kew list of Mr. Godman's plants, as if collected by him in Fayal; probably the present species being intended. The short, wide, doubly winged pods on my own specimens are quite decisive as to it being *sativus* in Fayal; where, however, I suspected it to be only an occasional straggler from cultivation And perhaps the two following also are in the same condition.

121. Lathyrus Clymenum, Linn. South Europe.

Isles 3. Miguel, Terceira, Fayal. Watson cat. 53. Hunt coll. Drouet cat. 154.

Under the name of *L. articulatus* in the two Catalogues cited, and doubtless labelled accordingly. I now believe it ought to be *Clymenum*; but much confusion has arisen between the two through mislabelled specimens from other habitats.

122. Lathyrus tingitanus, Linn. South Europe.

Isles 2. Miguel, Fayal. Seubert flo. 389. Watson cat. 51. Hunt coll. Introduced?

21. Rosaceæ.

123. Prunus lusitanica, Linn. Spain. Portugal.

Isle 1. San Miguel. Hunt coll. Godman coll. Drouet cat. 107. Native.

Prunus Cerasus is not a native of the Isles. M. Drouet errs in supposing that *Cerasus* intended the same species with *lusitanica*. By the comment in his Catalogue, page 86, he would seem to give me credit for not being able to distinguish between the "Morello Cherry" and the "Portugal Laurel," both so common in English gardens.

124. Spiræa Filipendula, Linn. Europe. Britain.

Isles 2. San Miguel, introduced; Hunt coll. Terceira (Hartung); Drouet cat. 106.

The reported existence of this plant also in Terceira leads to its inclusion in the present list; otherwise, on

Mr. Hunt's intimation of an alien origin, it would have been omitted here.

125. Rubus fruticosus, "Linn." Europe. Britain. Isles 5. Miguel, Maria, Terceira, Pico, Fayal. Seubert flo. 355. 45. Watson cat. 78. Hunt coll. Godman coll. Drouet cat. 95.

This is the segregate species usually named *discolor* by English botanists, at any rate, so far as the specimens seen by myself are in account.

126. Rubus Hochstetterorum, Seubert. Azores only. Isles 4. Miguel, Pico, Fayal, Flores. Seubert flo. 356. 44. Watson cat. 76. Hunt coll. Drouet cat. 96.

Mr. P. B. Webb held this "a large southern form of *R. fruticosus*." I should be surprised to get such an opinion from any Swedish botanist. In England, though not in Sweden, *R. discolor* is adopted for the special representative of *fruticosus*; and it was likely that English representative which Mr. Webb had in view. Still, *discolor* and *Hochstetterorum* are wide asunder in my own eyes; and countless descents must have passed since their common ancestor was one and the same thing on Darwinian views.

127. Fragaria vesca, Linn. Europe. Britain. Isles 4. Miguel, Pico, Fayal, Flores. Seubert flo. 357. Watson cat. 79. Hunt coll. Godman coll. Drouet cat. 97.

128. Potentilla anserina, Linn. Europe. Britain. Isles . . . . ? "Hab. ubique in pascuis siccioribus et ad vias;" Seubert flo. 360. "Tout l'archipel;" Drouet cat. 102; but not marked as a species actually seen.

The admitted existence of this plant in the Isles apparently rests solely on the testimony of Seubert's Flora, where no second number is added to show that specimens were actually distributed by Hochstetter. It is impossible, under the circumstances, not to suspect some gross error or careless statement on the part of Seubert's informant. A conspicuous plant, to be found "everywhere," would not have been wholly overlooked by all the other collectors, even by those who collected the commonest species, in order to obtain a full list of the insular flora.

129. POTENTILLA REPTANS, Linn. Europe. Britain.
Isles . . . . ? "Hab. hinc inde locis humidiusculis;" Seubert flo. 361. Pico; Drouet cat. 101.

I have not seen examples of this from the Isles; but some of the specimens referred by myself to the widely varying *Potentilla Tormentilla* (of authors) do bear so much of a first-glance resemblance to the present species, that I can easily conceive the name being entered in a list of plants supposed to have been seen by a traveller. It does not appear from Seubert's Flora that specimens were collected by Hochstetter for distribution, nor does Dr. Seubert give any intimation (direct or indirect) that specimens of it had been seen by himself. Not reported for the Canaries, and only very recently and locally for Madeira.

130. POTENTILLA TORMENTILLA, Auct., cum varr.
Europe. Britain.
Isles 5. Miguel, Terceira, Pico, Fayal, Flores. Seubert flo. 358 & 359. Watson cat. 80. Hunt coll. Godman coll. Drouet cat. 99 & 100.

The same difficulty is experienced here with the subspecies or varieties of the Linnean *Tormentilla*, which has been so frequently acknowledged by English and European botanists. The Isles produce all or most of the forms variously named by the botanists of Britain and of Europe, "*officinalis, erecta, reptans, procumbens, nemoralis, divergens, mixta,*" two or three of them sufficiently dissimilar from the rest in their extreme examples, but apparently so graduating into each other as to prevent any absolute diagnosis between them, while allowing nice opportunities to petty minds to make petty distinctions on paper.

131. POTENTILLA VERNA, Linn. Europe. Britain.

Isle 1. Santa Maria; Drouet cat. 98. Marked as having been actually seen; but not reported by other collectors.

We learn by Lowe's Manual Flora of Madeira, that the same species has been erroneously reported from that island. It cannot be accepted as a member of the Azoric flora until confirmed on higher botanical authority. Thus, out of four species here enumerated, I am prepared to accept one only as certain; that one divisible into other segregates or subspecies, however, according to the ideas of individual botanists.

132. ALCHEMILLA ARVENSIS, Scop. Europe. Britain.

Isles 2. San Miguel; Hunt coll. Fayal; Watson cat. 82. An inconspicuous weed easily overlooked.

133. AGRIMONIA EUPATORIA, Linn. Europe. Britain.

Isles 4, or more. Miguel, Terceira, Graciosa, Fayal. "Nearly all the islands;" Seubert flo. 362. 42. Watson cat. 81. Hunt coll. Godman coll. Drouet cat. 103.

134. POTERIUM SANGUISORBA, Linn. Europe. Britain.

Isle 1, or more. "In graminosis hinc inde;" Seubert flo. 363. San Miguel; Drouet cat. 104, but not marked as a species actually seen.

22. Crassulaceæ.

135. Tillæa muscosa, Linn. Europe. England. Isles 4. Miguel, Pico, Fayal, Flores. Seubert flo. 302. Watson cat. 92. Hunt coll. "Hinc inde;" Seubert.

136. Umbilicus pendulinus, DC. Europe. Britain. Isles 3, or more. Miguel, Fayal, Flores. Seubert flo. 303. Watson cat. 91. Hunt coll. Godman coll. Drouet cat. 196.

Drouet's Catalogue enumerates also *Umbilicus horizontalis* of De Candolle as a second species, observed on walls in Santa Maria. If I rightly understand that alleged species, it is not even a permanent variety, but simply a variation by place or season.

137. Aichryson villosum, Webb. Madeira. Canaries. Isle 1. Santa Maria; Hunt coll. (N.B. The Santa-Maria specimens of this and other plants were all received direct from Mr. Hunt; but I presume that they were actually collected there for Mr. Hunt by Mr. Schloss.)

I am not prepared to decide whether this is truly the *Sempervivum* to which the name *villosum* would be applied by the Rev. R. T. Lowe. Nor can I see how to distinguish between my Gomera specimen, from Bourgeau, no. 443, and a Madeiran specimen "*Sempervivum villosum*" by Dr. Lemann, except that the latter is decidedly less pubescent, which is just the contrary of what ought to be the case, according to Mr. Lowe's diagnosis.

23. Haloragiaceæ.

138. Myriophyllum alterniflorum, De Cand. Europe. Britain. Isle 1. San Miguel; T. C. Hunt coll. "Fayal

(Watson);" Drouet cat. 113; but some error in this latter record.

Two specimens are retained in my herbarium. The male flowers are clearly alternate on one of them; on the less advanced other specimen they seem to be verticillate. Having been distorted in drying, it is difficult now to say whether the spikes were originally erect or drooping. Thus, the name *alterniflorum* may be held not absolutely certain.

139. CALLITRICHE VERNA, Auct. Europe. Britain. Isles 4. Miguel, Pico, Fayal, Flores. Seubert flo. 166. Watson cat. 84. Hunt coll. Godman coll.

This is the plant which, in England, we now label "*platycarpa*;" whether it is equally the *C. stagnalis* (Scop.), I am not prepared to say. The Pico specimens only are in fruit, and they want the lower leaves.

24. MYRTACEÆ.

140. MYRTUS COMMUNIS, Linn. South Europe. Isles 2, or 3. Miguel, Maria, Fayal? Hunt coll. Godman coll. Drouet cat. 104. Fayal; Baron do Castello de Paiva, in Kew Herbarium,—wild?

Mr. Hunt deemed this clearly indigenous in the Isles, though now become very scarce through being gathered for use by the tanners.

25. LYTHRACEÆ.

141. PEPLIS PORTULA, Linn. Europe. Britain. Isles 3. Fayal and Pico; Watson cat. 88. Miguel; Hunt coll. and Godman coll. Perhaps rare, as apparently not observed by the German and French collectors.

142. Lythrum Hyssopifolia, Linn. Europe. England. Isles 7. Miguel, Maria, Terceira, Pico, Fayal, Flores, Corvo. Seubert flo. 354. 41. Watson cat. 87. Hunt coll. Godman coll. Drouet cat. 94.

143. Lythrum Græfferi, Tenore. South Europe. Isles 4. Terceira; Seubert flo. 353. 40. San Jorge; Hunt coll. Pico; Baron do Castello de Paiva in Kew Herbarium. Terceira; Morelet. Santa Maria (Hartung); Drouet cat. 93.

It is to be feared there was some confusion in the labeling of my original specimens (no. 86, 1842), leaving it now uncertain whether any of them truly belonged to the present species, or all of them to *Hyssopifolia*. On this account I give the authorities on which the several islands are cited.

26. Onagraceæ.

144. Epilobium parviflorum, Schreber. Eur. Britain. Isles 2. San Miguel; T. C. Hunt coll. Flores; Watson cat. 83. Drouet cat. 110.

Drouet's Catalogue enumerates also two species of *Œnothera,* doubtless casual escapes from cultivation.

27. Cucurbitaceæ.

145. Ecbalium Elaterium, Rich. South Europe. Isle 1. Santa Maria; Hunt coll. and Godman coll. Not San Miguel, to which Mr. Hunt's specimens are incorrectly assigned in Drouet cat. p. 87.

28. Umbelliferæ.

146. Sanicula azorica, Guthnick. Azores only. Isles 5. Miguel, Maria, Terceira, Pico, Fayal. Seubert

flo. 293. 24. Watson cat. 95. Hunt coll. Drouet cat. 169.

A remarkable plant, quite distinct from the European species. Well-named *S. ciliata* by Solander in the Banksian herbarium, altered to *S. ciliaris* in the herbarium of J. E. Smith.

147. Conium maculatum, Linn. Europe. Britain. Isles 2. San Miguel; Hunt coll. Santa Maria; Godman coll. Introduced?

148. Smyrnium Olusatrum, Linn. Europe. Britain? Isles 2. San Miguel, in one place only; Hunt coll. Fayal; Drouet cat. 188; but not marked as a species actually seen in the Isles, and how the author gets the habitat of Fayal I am at a loss to explain.

149. Apium graveolens, Linn. Europe. Britain. Isles 6. Miguel, Terceira, Graciosa, Pico, Fayal, Flores. Watson cat. 101. Hunt coll. Godman coll. Drouet cat. 172. Terceira, subspontanea; Baron do Castello de Paiva in Kew Herbarium.

150. Helosciadium nodiflorum, Koch. Europe. Britain. Isles 2. Terceira; Seubert flo. 295, and Drouet cat. 173. Miguel; Hunt coll. Azores; Godman coll.

This appears to be a variable species everywhere. Mr. Godman's specimen is just one of the ordinary small creeping states, with short peduncles, often labelled "*repens*" in England. Mr. Hunt's example is larger, and with longer peduncles.

151. Ammi Visnaga, Lam. South Europe. Isles 2. San Miguel; Hunt coll. and Godman coll. Santa Maria; Drouet cat. 176.

152. Ammi Huntii, Watson. Azores only.

Isle 1. San Miguel; Hunt coll. 1846. Pico (Hartung); Drouet cat. 175. (*A. procerum* of Lowe?)

This fine species is described in my former Supplementary Notes; that is, in Hooker's Journal of Botany, vol. vi. p. 384, and in the London Phytologist for July 1847. It can hardly be the same with Mr. Lowe's *Ammi procerum*.

153. Ammi majus, Linn. South Europe.

Isles 2. Graciosa, Flores; Drouet cat. 174,—apparently the sole authority for this species in the Isles.

Ammi majus is stated to grow in Madeira and the Canaries, so far giving a sort of presumption in favour of it being really found in the Azores also. But there is another undescribed umbelliferous plant in the small island of Flores, seemingly unoticed by M. Drouet, which much resembles the *Ammi majus*, and which might be mistaken for this latter, if the differences between the involucral leaves were not recognized; neither does it appear that M. Drouet had himself seen *Ammi Huntii*.

154. Petroselinum trifoliatum, Watson. Azores only.

Isle 1. On rocks near Santa Cruz, in Flores; Watson cat. 103. Seen in the one spot only, and by no other collector.

According to Bentham and Hooker (Genera Plantarum, vol. i. p. 889) this and the next species would be better placed in the genus *Ammi*, notwithstanding their undivided involucral bracts. This plant would not have been mistaken for the *Ammi majus* by M. Drouet by reason of the undivided bracts and other characters. But, as above remarked, there is still another allied umbelliferous plant in Flores which is considerably more like to *Ammi majus*. I possess only a single example of that undescribed species, insufficient to afford a proper diagnosis. In leaves and general appearance it might be held an intermediate species between *majus* and *Huntii*; but instead of the trifid or subpinnatifid involucral bracts of these two species, those of my solitary specimen are all quite entire and linear-lanceolate, very like those of *Petroselinum trifoliatum*. Indeed, notwithstanding the dissimilar leaves, I had at first confused it with the *P. trifoliatum*; but after raising this latter in England from seed collected in Flores, I was led to regard them as two different species.

155. Petroselinum Seubertianum, Watson.

Azores only.

Isles 2. San Miguel; Hunt coll. Pico; Watson cat. 100, and Supplementary Notes, p. 387. *Kundmannia sicula*, Seubert Flora Azorica, no. 298, but erroneously.

This is certainly not the true *Kundmannia sicula* (*Brignolia pastinacæfolia*), although there is some considerable resemblance between them in foliage. As with the preceding species, Bentham and Hooker refer this plant also to the genus *Ammi*.

156. Petroselinum sativum, Hoffm. South Europe.

Isle 1, or more. "Hinc inde prope littora maris;" Seubert flo. 294. Miguel; Hunt coll. Godman coll. "Hab. tout l'archipel;" Drouet cat. 170.

157. Pimpinella villosa, Schousboe. Portugal. Spain.

Isle 1. San Miguel; T. C. Hunt coll. Apparently not found by any other collector.

According to the Prodromus of De Candolle, this is distinguished from *P. bubonoides* by its villose petals; but the specimens from Spain and Portugal (no. 1884) distributed by Bourgeau, labelled with the name of *P. bubonoides*, have the same villose petals with the Azore plant, and otherwise differ only slightly by their rounder leaves less cuneate downwards. If the Lusitanian specimens truly represent *bubonoides*, then that and *villosa* may likely prove to be only a single species. I took the latter name from a label in the herbarium of Sir William Hooker.

158. PIMPINELLA DICHOTOMA, Linn. Spain. Africa.

Isle 1. "In apricis prope litus insulæ, Pico;" Seubert flo. 296. Pico; Drouet cat. 177.

This habitat is in some degree confirmed by being repeated in Drouet's Catalogue with the usual mark prefixed to the name of the plant, indicating that it had been seen by himself or companion. See *Crithmum* below for ground of doubt.

159. CHÆROPHYLLUM AROMATICUM, Linn., var.? East Europe.

Isles 2. Flores, on the north side, near San Pedro; Watson cat. 102. Miguel; Hunt coll.

This name is given in some uncertainty. The specimens are much more pubescent than usual with *C. aromaticum*, and the leaves are irregularly inciso-serrate, giving to the plant somewhat the aspect of *Myrrhis odorata*. The Azore Isles would be an unexpected habitat for *C. aromaticum*; but I feel very doubtful of the nativity of the plant in its one locality in Flores.

160. FŒNICULUM VULGARE, Gærtn. Europe. Britain.

Isles 3. Miguel, Fayal, Flores. Seubert flo. 297. Watson cat. 97. Hunt coll. Godman coll. Drouet cat. 179.

161. CRITHMUM MARITIMUM, Linn. Europe. Britain.

Isles 5. Miguel, Pico, Fayal, Flores, Corvo. Watson cat. 96. Hunt coll. Godman coll. Drouet cat. 180.

It is remarkable that the *Crithmum* should have been omitted from Seubert's Flora, although found in so many of the Isles, and by all the other collectors following Hochstetter. Is it possible that the "*Pimpinella dichotoma*" of the Flora Azorica really intends the *Crithmum*? See the former above.

162. Angelica montana, Schl. N. M. Europe.

Isle 1. Terceira, in the Caldeira; Drouet cat. 181. Not reported by any other collector. Accompanied by a remark in the work cited that M. Heer deems it *Angelica sylvestris*, or a variety of this latter.

163. Coriandrum sativum, Linn. East Europe.

Isle 1, or 2. Miguel; Hunt coll. Terceira; Drouet cat. 187, not marked as a species actually seen. Introduced?

164. Daucus Carota, Linn., var. Europe. Britain.

Isles 5. Miguel, Terceira, Graciosa, Fayal, Flores. Seubert flo. 299. Watson coll. 99, and Supplementary Notes, p. 386. Hunt coll. Godman coll. Drouet cat. 183 and 185.

Several varieties occur in the Isles, which would be held for species by botanists addicted to segregating species; but which I am not prepared to refer positively to any thing but the aggregate *Daucus Carota* of Linnæus and the older authors. According to the late Dr. Lemann, some of my specimens belong to *D. neglectus* of Lowe, in which as yet I see only a weakly state of *D. Carota*. Others from Mr. Hunt might represent our sea-side form, the *maritimus* of Withering, and *gummifer* of later writers. In Flora Azorica, the species from Terceira is named *polygamus*, Gouan.

165. Torilis tenuifolia, Lowe? Madeira? Canaries? Isles 4. Miguel, Maria, Graciosa, Fayal. Seubert flo. 300. Watson cat. 98. Hunt coll. Godman coll. Drouet cat. 189.

This is the *T. helvetica* of Seubert's Flora and Drouet's Catalogue, the *T. infesta* of my own collection. It differs from our British *infesta* by its less rigid and more divided leaves, and by its less rough and more slender stem and branches. In Dr. Lemann's Madeira list the only two species enumerated are *nodosa* and *infesta*; the latter subsequently divided by Mr. Lowe into three several species, all alleged to be distinct from the European *infesta* or *helvetica*.

N.B. *Melanoselinum decipiens*, of Hoffman, may be rejected from the Azore lists. The two leaves from the young plant in the Caldeira of Fayal, which were mentioned in my own former catalogue, can hardly have belonged to so large a species as this is described to be. Unfortunately I have seen only the flowering branchlets of the *Melanoselinum* from Madeira, without the lower leaves, or those of the young plant.

29. Hederaceæ.

166. Hedera canariensis, Willd. Canaries. Isles 4. Miguel, Pico, Fayal, Flores. Seubert flo. 301. Watson cat. 93. Hunt coll. Godman coll. Drouet cat. 192.

Seubert enumerates the Azore Ivy in his Flora simply as the European *Helix*. Drouet adds "var. *hibernica*" to the same specific name, as had been done also in my own former catalogue, on recollection and on inspection of the merest fragment without flowers or fruit. Specimens with the berries and floral leaves, which have been obtained through the kindness of Mr. Hunt and Mr. Godman, now suggest that the Azore Ivy is really nearer to the *canariensis* from Mr. Webb than to our garden *hibernica*, although in some sort an intermediate form between these two. In my three Canary specimens, the leaves immediately below the flowering panicles are decidedly cordate, that is, very broadly ovate in form and deeply lobed at the insertion of the petiole. In the Azore specimens they are still cordate, though with a rather less deep sinus at the base. In the variety *hibernica* of our gardens, the leaves are simply rounded at the base, or almost straight transversely, being broadly ovate, or triangularly ovate, but not truly cordate; in the ordinary *Helix* they usually narrow downwards to the petiole, being cuneate-ovate or ovate-lanceolate. I have seen no stem-leaves from the Canaries or Azores which were stellately or angularly lobed in the form usual with the wild *Helix* of Britain and Europe. Still, as described by Webb, they are either entire or slightly 3–5-lobed upwards; and as there is a wide range of variations among the leaves of the ordinary *Helix* of Britain, such also may be the case to some extent with those of the Azores and Canaries. A long series of the leaves of the British wild *Helix* has been collected into my herbarium, for the purpose of learning their range of variation in form; but I have never seen a leaf of it which the eye would not readily and instantly distinguish from those of the Azores and Canaries; while their thickness and texture is so different from the latter, that I could sort them blindfold by the touch alone. The Ivy of Madeira I do not know. In his Manual Flora of that island, Mr. Lowe treats the Madeira Ivy as ordinary *Helix*, not even designating it as a variety, although he refers to Webb and Berthelot's Phytographia Canariensis. This is truly remarkable, because Mr. Lowe usually appears to take an especial pleasure in describing small differences, and in founding new species and named varieties on characters which many botanists might pass over as scarcely more than trifling variations or individual physiognomies. I do not at all pretend to possess the like taste for little distinctions,—useful as it is in local describers; but still I must hold that the Ivy of the Canaries and Azores is sufficiently different from the *Helix* of Europe and Britain to require a varietal name, if not also to justify a distinct specific name.

The so-called "Irish Ivy" (*hibernica* of the gardens) is not known in a wild state in Ireland; it stands between *Helix* and *canariensis,* corresponding with the former in the outline or shape of its leaves, but with the latter in their much larger size and coriaceous thickness. The "Palmate Ivy" of the gardens is nearer to *Helix* than to *canariensis* in the texture, size, thickness, and venation of its leaves; while it comes closer to *canariensis* in the forward or upward direction of their lobes. But on the species of *Hedera* the views of Dr. Seemann should be consulted, in 'Journal of Botany,' volume second, pages 304–7, also volume third, pages 201–3; they refer the Ivies of Madeira and Portugal to *canariensis*; also *hibernica* and *palmata.*

30. Caprifoliaceæ.

167. Viburnum Tinus, Linn. South Europe.

Isles 4. Miguel, Fayal, Flores, Corvo. Seubert flo. 239. Watson cat. 105. Hunt coll. Godman coll. Drouet cat. 194.

The leaves on my herbarium specimens from the Isles are broader and more obtuse than those of the European examples or of our common garden "Laurestinus." Dr.

Seubert refers his specimens to the *Viburnum lucidum* of Miller, but simply as a variety of *Tinus*.

168. Sambucus nigra, Linn. Europe. Britain.
Isles 3. Fayal, Flores; Watson cat. 104, "aliena." Miguel; Hunt coll. "Hab. toutes les îles dans les haies;" Drouet flo. 193, but not marked as seen by himself or companion. May it not be suggested that Drouet's indication is rather European than Azoric? Stone walls and banks topped by rows of *Arundo Donax* would hardly be suitable "hedges" for the elder-bush.

31. Rubiaceæ.

169. Rubia splendens, Hoffmansegg et Link. Portugal?
Isles 4. Miguel, Pico, Fayal, Flores. Seubert flo. 237. 45. Watson cat. 106. Hunt coll. Godman coll. Drouet cat. 268.

The name of *splendens* is adopted from the Flora Azorica, although the midrib of the leaf in my specimens is quite smooth on the upper surface, aculeated only underneath. The *R. angustifolia* of Lowe's Flora of Madeira comes between the Mediterranean *Requieni* (same as *peregrina* of Bourgeau's Pl. Can. no. 278) and the Azoric *splendens*; and yet these two latter look widely dissimilar, if compared with each other apart from the Madeira specimens. I find difficulty in believing the Azore plant a form of *Rubia peregrina*.

170. Galium Mollugo, Linn. Europe. Britain.
Isles .....? "Hab. In graminosis, ad vias;" Seubert flo. 233. "Hab. tout l'archipel;" Drouet cat. 263; the name being printed without the mark used to show it that of a species actually seen.

It is to be feared that we have here another example of

the very loose manner in which plants have been admitted to swell the lists of species in the two books cited. Mr. Hunt collected every kind of fern and flowering plant that he found in San Miguel (the largest of the islands); and he was a resident. I did the like in four other of the islands, as far as my opportunities enabled this to be done. And it is shown by their collection that Mr. Godman and his assistant were active collectors of plants, although botany was not the primary object with them. Yet this conspicuous *Galium* was not collected by any one of the four Englishmen, nor seen by either of the two Frenchmen. Under these circumstances it is incredible that the *Galium* should occur in frequency to warrant the vaguely general habitat indicated by Seubert, or the positively universal one stated by Drouet. Is it not more likely that *Galium mollugo* is absent from all the Isles, rather than present along road-sides in all of them? It is wanting in Madeira and Canaries. On the other hand, there is difficulty in suggesting any constituent in the Azore flora which could be mistaken for *Galium mollugo* by any botanist.

171. GALIUM PALUSTRE, Linn. Europe. Britain. Isles 3. Miguel, Fayal, Flores. Seubert flo. 234. Watson cat. 109. Hunt coll. Drouet cat. 264 & 265.

Dr. Seubert likens his Flores specimens to *Galium debile*, "quod mera *Galii palustris* videtur varietas." M. Drouet pounces upon this remark to swell the count of his list by the addition of "*Galium debile*, Hoffm. et Link," as well as "*Galium palustre*," both marked as actually seen in Flores. My own specimens from Flores are smooth.

172. GALIUM ANGLICUM, Huds. Europe. England. Isles 6. Miguel, Terceira, Pico, Fayal, Flores, Corvo.

Seubert flo. 236. 6. Watson cat. 110. Hunt coll. Godman coll. Drouet cat. 266. Terceira; Baron do Castello de Paiva, in Kew Herbarium. Corvo (Hartung); Drouet flo. Generally admitted to be a form of *Galium parisiense.*

173. Galium Aparine, Linn. Europe. Britain. Isles 2, or more. Miguel, Fayal. Seubert flo. 235. Watson cat. 108. Hunt coll. Godman coll. Drouet cat. 267.

174. Aspera muralis, Lowe. South Europe. Isle 1. Miguel; T. C. Hunt coll. 1846. *Galium murale,* De Cand. prod. *Sherardia muralis,* Linn. Strangely enough, Mr. Hunt's examples of this plant somehow got mingled with those of *Galium anglicum,* and possibly some of them may have been so distributed, accompanied by the labels of *Galium anglicum.*

175. Sherardia arvensis, Linn. Europe. Britain. Isles 4. Miguel, Terceira, Fayal, Flores. Seubert flo. 238. Watson cat. 107. Hunt coll. Godman coll. Drouet cat. 269.

32. Valerianaceæ.

176. Valerianella dentata, Koch. Europe. Britain. Isle 1. Pico; Watson cat. 111. I no longer remember the conditions under which this plant was found in Pico. No other collector has reported it in the Isles, though a likely plant to have been introduced as a weed of agriculture.

33. Dipsaceæ.

177. Scabiosa nitens, Röm. et Schultz. South Europe. Isles 5. Miguel, Terceira, Fayal, Flores, Corvo. Seu-

bert flo. 194 & 195? Watson cat. 112. Hunt coll. Godman coll. Drouet cat. 200–204.

A difficult plant. The name is repeated from the Flora and Catalogues cited, as probably the name belonging to the species which Seubert makes into two, and Drouet into five. The grounds for selecting the name are more presumptive than strictly technical. First, because the Azore Isles are recorded as the habitat of *S. nitens*; second, because I must refer to a single species all the specimens seen by myself, which show it from three of the five islands named; third, because the imperfect description of *nitens*, so far as it goes, agrees passably well with some of the specimens, though not well adapted to the whole of them. The seven specimens now in my herbarium range in size from 3 to 18 inches, and the foliage varies in form and cutting with the luxuriance of the plants, the smallest having the leaves spathulate-elliptic, coarsely serrate, but otherwise undivided, the larger examples having their uppermost leaves lyrate-pinnatifid, the intermediate leaves showing intermediate forms and cutting. Perhaps my most similar herbarium specimen to those from the Azores is one from Madeira, labelled "*Scabiosa atropurpurea*" by Dr. Lemann, and which is supposed to be *S. maritima*, var. *pallidiflora*, of Lowe's Flora. I can scarcely conceive that Madeira specimen belonging to the *Scabiosa atropurpurea* of our gardens, although Mr. Lowe apparently holds the former to be a "more truly wild" state of the latter. In outline form the leaves of the Azore specimens come even nearer to *S. atropurpurea* of English gardens than do those of that one Madeira specimen; but their venation is reticulately ramified, their surface smooth and shining, and their substance thicker, almost leathery coriaceous. In none of my Azore specimens can the upper leaves be described by Mr.

Lowe's words, "upper pectinately-pinnatipartite," while this combination of terms quite fits the stem-leaves of the one Madeira specimen from Dr. Lemann. The specific name of *nitens* is retained for the present, as one truly belonging to the Azore plant; although it may hereafter appear that the plant itself is a local form of *Scabiosa maritima*.

34. Compositæ.

178. Galactites tomentosa, Mœnch. South Europe. Isles 3, or more. Miguel, Pico, Fayal. "All the islands;" Seubert flo. 215. 109. Watson cat. 185. Godman coll. Drouet cat. 234.

179. Cirsium lanceolatum, Scop. Europe. Britain. Isle 1. Pico. Seubert flo. 217. Watson cat. 187, under the generic name of *Carduus*.

180. Carduus pycnocephalus, Linn.? Europe. Britain? Isles 2, or more. Miguel, Pico. "All the islands;" Seubert flo. 216. Watson cat. 186. Hunt coll. Drouet cat. 255.

Carduus tenuiflorus of Curtis and Smith, according to Seubert and to Drouet; but the plant known by that name in England is not exactly the Azore plant. Mr. Bentham, in Handbook of the British Flora, adopts the name *pycnocephalus*, "Jacq.," for the British plant. Mr. Lowe cites the Floras of Seubert and of Bentham for the Madeira species named by him *tenuiflorus*, not *pycnocephalus*, Linn.

181. Centaurea melitensis, Linn. South Europe. Isles 3. Maria, Fayal, Pico. Pico; Seubert flo. 214. 110. Fayal; Watson cat. 184. Maria; Hunt coll.

182. Erigeron canadensis, Linn. America. Europe? Isles 5. Miguel, Terceira, Fayal, Flores, Corvo. Watson cat. 193. Hunt coll. Godman coll. Drouet cat. 207.

This plant appeared to be as well established in the Isles as any other way-side weed. It would thus be held native in the Azores, if its name and a traditionary belief did not attribute its origin to America. Curious that it found no mention in the Flora Azorica; possibly there confused with the next plant?

183. Conyza ambigua, De Cand. South Europe. Isles 3 or more. Miguel, Fayal, Flores. "All the islands;" Seubert flo. 198. 108. Watson cat. 194. Hunt coll. Godman coll. Drouet cat. 208.

184. Solidago azorica, Hochst. Azores only. Isles 6. Miguel, Terceira, Pico, Fayal, Flores, Corvo. Seubert flo. 197. 106. Watson cat. 192. Hunt coll. Drouet cat. 206.

185. Bellis perennis, Linn. Europe. Britain. Isle 1. Terceira; Godman coll. Is it in the Kew Herbarium from Guthnick's collection?

186. Seubertia azorica, Watson. Azores only. Isles 6. Miguel, Terceira, Pico, Fayal, Flores, Corvo. Seubert flo. 196. 101. Watson cat. 202. Hunt coll. Drouet cat. 205.

The figures of this Daisy in the Flora Azorica, while they give a fair general idea of the plant, yet fail in the most important details. The lower figure represents a luxuriant plant in the early flowering stage; the *flat* receptacle and the *reflexed* involucral bracts of the anthodium are not at all shown in the figure. These are two of the characters which distinguish the genus from *Bellis*, while the general habit is so like that of our *Bellis perennis*, that it may perhaps be thought better to place *Seubertia* as a subgeneric section. The character "Achenia in verrucas squamæformes insidentia" was taken from dried specimens, and may not be literally exact, though the surface of the flat receptacle is manifestly uneven. An entirely different Spanish plant has been distributed, under the name (interrogatively?) of "*Bellis azorica*, Hochst." It was substituted for *Antirrhinum majus*, no. 33, of Dr. Moritz Willkomm's Spanish plants. My example of this is in flower only; but it is apparently a true *Bellis*, likely the *B. dentata* of De Candolle, and quite a distinct thing from the *Seubertia azorica*. This generic name had been earlier applied to a genus of *Liliaceæ*; and if it thus cannot stand for the *Bellis azorica*, the change to *Vidalia azorica* may be suggested as an appropriate one (see page 188).

187. Chrysanthemum Myconis, Linn. South Europe. Isles 5. Miguel, Maria, Terceira, Pico, Fayal. Seubert flo. 203. Watson cat. 197. Hunt coll. Godman coll. Drouet cat. 217. Terceira; Baron do Castello de Paiva, in Kew Herbarium. *Pyrethrum Myconis* of Drouet cat.

188. Chrysanthemum segetum, Linn. Europe. Britain. Isles 5. Miguel, Maria, Pico, Fayal, Flores. Seubert flo. 204. Watson cat. 198 & 199. Hunt coll. Godman coll. Drouet cat. 218.

189. Chrysanthemum coronarium, Linn. S. Europe. Isles 3. Miguel, Terceira, Fayal. Seubert flo. 205. Watson cat. 0. Hunt coll. Godman coll. Drouet cat. 219.

Although generally thus reported by the collectors, this

seems to be clearly an alien, casually wild in the Isles. The European *Chrysanthemum Parthenium* (the *Matricaria Parthenium,* Linn.) was sent from San Miguel by Mr. Hunt. And the *Chrysanthemum pinnatifidum,* of Madeira is recorded from the same island by Seubert and by Drouet; but Mr. Hunt informed me that it had occurred to himself only in or near gardens.

190. ANTHEMIS COTULA, Linn. Europe. Britain.

Isles 4. Miguel, Terceira, Fayal, Flores. Seubert flo. 200. Watson cat. 196. Hunt coll. Godman coll. Drouet cat. 211.

Drouet's Catalogue enumerates both this species and its near ally *A. arvensis*; giving San Miguel and Pico as habitats for the latter, with the usual bold guess "tout l'archipel" for the former, and marking both as species actually seen.

191. ANTHEMIS AUREA, De Cand. Portugal. Madeira.

Isles 6. Miguel, Maria, Terceira, Pico, Fayal, Flores. Seubert flo. 201. 100. Watson cat. 195. Hunt coll. Godman coll. Drouet cat. 212.

In the Manual Flora of Madeira, Mr. Lowe explains the differences between this species and the *Anthemis nobilis.* He considers it the *Anacyclus aureus* of Brotero, but not of Lamarck.

192. ACHILLEA MILLEFOLIUM, Linn. Europe. Britain.

Isle 1. San Miguel; F. D. Godman coll. Apparently not found by any other collector, though here, as in the case of other very common European species, the chance of it having been seen and neglected offers an alternative explanation. According to Mr. Lowe, it was introduced to Madeira half a century ago, and is now used there in decoctions and fomentations; perhaps such may be the case in San Miguel also.

193. Gnaphalium luteo-album, Linn. South Europe. Isles 6. Miguel, Terceira, Graciosa, Pico, Fayal, Flores. Seubert flo. 207. 104. Watson cat. 189. Hunt coll. Godman coll. Drouet cat. 221.

The American *Gnaphalium pennsylvanicum* was recorded in Seubert's Flora as found in Terceira and Fayal. It was collected by Mr. Godman in the former island.

194. Filago germanica, Linn. Europe. Britain. Isles 3. Miguel, Fayal, Flores. Seubert flo. 209. 106. Watson cat. 190. Hunt coll. Godman coll. Drouet cat. 223.

This is the common British form, the *F. canescens* of Jordan. The other segregate species should be looked for.

195. Filago gallica, Linn. Europe. England. Isles 2. Miguel, Fayal. Seubert flo. 210. Watson cat. 291. Hunt coll. Drouet cat. 224.

196. Senecio vulgaris, Linn. Europe. Britain. Isles 3. Miguel, Terceira, Fayal. Watson cat. 201. Hunt coll. Godman coll. Perhaps disregarded by the French and German collectors.

197. Senecio sylvaticus, Linn. Europe. England. Isle 1. Santa Maria; Hunt coll. Treated as a wild plant of Madeira in Lowe's Manual Flora.

198. Senecio erraticus, Bert. South Europe. Isle 1. San Miguel; Hunt coll. Likewise sent from the same island by Mr. Schloss to Mr. Sansom.

199. Senecio malvæfolius, De Cand. Azores only? Isles 4. Miguel, Maria, Terceira, Fayal. Seubert flo. 212. 112. Watson cat. 200 "*maderensis*." Hunt coll. Drouet cat. 228. Terceira; Baron do Castello de Paiva, in Kew Herbarium.

This looks very near to *S. maderensis* on putting the herbarium specimens of the two side by side. Including two non-flowering stems, my own herbarium has half a dozen specimens from Madeira, a dozen from the Azores; the latter chiefly sent from San Miguel by Mr. Hunt. The leaves of the Azore plants are generally larger and less angular, and their petioles are more frequently without auricles at their base, and the flowers are more numerous and rather smaller. I detect no other marked differences; and I find that Mr. Lowe's diagnostic character of *S. maderensis* applies equally well to the Azore specimens, unless the two words "stem shrubby" be an exception. Dr. C. H. Schultz (bipontinus) names the San-Miguel plants "*Doronicum malvæfolium*" and "*D. malvæfolium*, var. *Huntii*," intending the auricled specimens as the variety. This variety was erroneously given as *maderensis* in my own former Catalogue when I had seen leaves only, without flowers.

The Flora Azorica records *Senecio pseudo-elegans*, a South-African species, as having become established in a maritime locality in San Miguel. Mr. Hunt found the same plant in San Miguel and Terceira, but only in gardens or evidently escaped from them.

200. Bidens leucantha, Willd. Madeira. Canaries. Isles 3. Miguel, Pico, Fayal. Seubert flo. 199. 113. Watson cat. 118. Hunt coll. Godman coll. Drouet cat. 209.

A form or variety of *Bidens pilosa* of Linnæus, and

likely an introduced plant in the Azores. Perhaps introduced also in Madeira and Canaries.

201. Calendula arvensis, Linn. South Europe.
Isles 3. Miguel, Terceira, Fayal. Watson cat. 203. Hunt coll. Godman coll.

The Flora Azorica enumerates only the garden *C. officinalis,* omitting the way-side weed *C. arvensis.* Drouet and his friends saw neither of them, unless their mark has been accidentally omitted before the specific names.

202. Xanthium Strumarium, Linn. var. S. Europe.
Isles 2. Miguel; Hunt coll. Flores; Watson cat. 204. Apparently rare or local.

In their fruiting stage the heads of the Azore plants are much thicker in proportion to their length than I have seen to be the case with any European example of *Strumarium;* they are almost globose, with beaks slightly bent.

203. Xanthium spinosum, Linn. South Europe.
Isle 1. Sent by Mr. Hunt from San Miguel. Not found by the other collectors.

204. Cichorium Intybus, Linn. Europe. Britain.
Isles 3, or more. Miguel, Fayal, Flores. Occasionally; Seubert flo. 218. Watson cat. 170. Hunt coll.

205. Tolpis nobilis, Hochst. Azores only.
Isles 6. Miguel, Maria, Terceira, Pico, Fayal, Flores. Seubert flo. 222. Watson cat. 182. Hunt coll. Godman coll. Drouet cat. 243. (Including also *Tolpis macrorhiza* of the Azore Catalogues, not that of Madeira.)

This is truly a wide-varying species, which possibly may be divisible into a group or series of segregate species. The figure of *Tolpis nobilis* in the Flora Azorica represents one of the extreme forms in regard of the shape and dentation of the leaves, their insertion on the stem, and the size of the flowers. Other forms occur which would be more nearly represented by the figure of *Tolpis macrorhiza* in the Botanical Magazine, no. 2988. Still, with Mr. Lowe's aid, I now consider that the true *macrorhiza* of Madeira has not occurred in the Azores, the varieties in these Isles often approximating to the Madeira species, some in one character, some in another character, but always retaining appreciable distinctions from the plants of Madeira.

206. Tolpis fruticosa, Schrank. Madeira. Canaries. Isles 3. Terceira, Fayal, Flores. Seubert flo. 221. 93. Watson cat. 179. Godman coll. Drouet cat. 242.

Aiton's name *succulenta* is a more appropriate one for the Azore plant than *fruticosa*. The leaves are succulent and vary from spathulate-obovate almost entire to linear-lanceolate and inciso-serrate. Still more dissimilar as this species is from the figure of *Tolpis nobilis* above referred to, there are some specimens in my herbarium not readily assigned between the two species, so connected are they with both by a series of transition forms. These are herbarium specimens, however, flowering stems or branches only. In the living plants the roots and mode of growth would remove all difficulty. As to naming formally all these different shapes, it appears to my judgment to be useless pedantry, so gradually do they pass into each other.

207. Tolpis barbata, Gaertner. South Europe.
Isles 3, or more. Several of the islands; Seubert flo. 219. Terceira; Godman coll. Maria, Flores; Drouet cat. 240.

208. Tolpis umbellata, Bert. South Europe.
Isles 4. Miguel, Terceira, Fayal, Flores. Seubert flo. 220, 91. Watson cat. 180. Hunt coll. Godman coll.

Under the name of *Tolpis crinita*, Lowe, in the Flora of Seubert; as two species, under both names, in the Catalogue of Drouet. This and the preceding are much alike, and possibly may be forms of a single species; but the far handsomer flowers of the *barbata*, the Hawkweed of our gardens, is a conspicuous difference, if not a very good technical distinction.

209. Thrincia nudicaulis, Lowe. Madeira. Canaries?
Isles 2, or more: Miguel, Fayal. Seubert flo. 223. Watson cat. 178. Hunt coll. Drouet cat. 224.

The Flora Azorica gives a figure of this plant; but no islands are named as its habitat. Mr. Hunt sent examples from San Miguel which quite correspond with Seubert's figure in the long beak of the fruit, and with them are other examples having fruit scarcely at all beaked. The Fayal plants come between these opposite extremes, as also do some of those from San Miguel. I presume that *Thrincia hirta*, no. 575 of Bourgeau's Plantæ Canarienses, must go to *T. nudicaulis*.

210. Thrincia hirta, De Cand. Europe. Britain.
Isles 5. Miguel; Hunt coll. Terceira; Barão do Castello de Paiva in Kew Herbarium? Terceira, Graciosa, Fayal; Drouet cat. 245. Flores; Dr. Mackay?

The Flora Azorica omits this, giving *T. nudicaulis* alone; but while I detect no other difference, excepting the length of the beak of the fruit, Mr. Hunt's specimens from San Miguel seem fairly divisible between the two species, real or supposed, in respect of that single character. M. Drouet adds *Thrincia hispida* (Roth) as a third species, all three actually seen in the Isles. This last is unknown to me, if any thing really distinct from the other two.

211. Helminthia echioides, Gaertn. Eur. England. Isles 4. Miguel, Terceira, Fayal, Flores. Seubert flo. 229. 95. Watson cat. 174. Hunt coll. Drouet cat. 255.

212. Urospermum picroides, Desf. South Europe. Isles 2, or more. Most of the islands; Seubert flo. 225. 96. Fayal; Watson cat. 175. Miguel; Hunt coll.

213. Hypochœris glabra, Linn. var. Eur. Britain. Isles 3, or more. All the islands; Seubert flo. 224. 90. Fayal, Flores; Watson cat. Miguel; Hunt coll. Drouet cat. 250.

Varies in being more or less pilose below, with scattered distant hairs.

214. Taraxacum officinale, Vill. Europe. Britain. Isles 2. Miguel, Fayal; Watson cat. 176. Hunt coll. Godman coll.

215. Lactuca Scariola, Linn. Europe. England. Isles 2. Fayal and Flores; Watson cat. 171. Not reported by any other collector; thus probably rare.

Another *Lactuca,* of uncertain specific name, is mentioned by Drouet as inhabiting the Caldeira in Fayal. This may be the "large-leaved *Composita*" mentioned at the end of my former Catalogue.

216. SONCHUS OLERACEUS, Linn. Europe. Britain. Isles 3. Miguel, Terceira, Fayal. Watson cat. 172. Hunt coll. Godman coll. Drouet cat. 256.

217. SONCHUS ASPER, Hoffm. Europe. Britain. Isles 3, or more. Miguel, Pico, Fayal. Seubert flo. 230. Watson cat. 173. Hunt coll. Drouet cat. 257. "*S. fallax,* Wallr.," of Seubert's flora.

218. CREPIS VIRENS, Linn. Europe. Britain. Isles 5. Miguel, Terceira, Graciosa, Fayal, Flores. Seubert flo. 231. 89. Watson cat. 181. Hunt coll. Drouet cat. 258, 259, 260.

This is made into three species in Drouet's Catalogue, through the enumeration of three synonyms as so many different species, "*polymorpha, virens, diffusa.*"

219. MICRODERIS RIGENS, De Cand. Azores only. Isles 4. Miguel, Pico, Fayal, Flores. Seubert flo. 227. Watson cat. 183. Hunt coll. Drouet cat. 252 and 253.

This is made into two species in Drouet's Catalogue by first enumerating it as a non-observed species, under the name of *M. rigens* of De Candolle, and then adding *M. umbellata* of Hochstetter as a second actually seen species; but it appears sufficiently clear that these two names belong to the same single species, as M. Drouet might have readily seen from my own former Catalogue, where the *umbellata* is quoted as a synonym under *rigens*.

220. Microderis filii, Hochst. Azores only.

Isles 3. Miguel, Fayal, Flores. Seubert flo. 228. Watson cat. 183. Hunt coll. Godman coll. Drouet cat. 254. Perhaps a variety of the preceding; somewhat intermediate forms being seen occasionally.

35. Campanulaceæ.

221. Campanula Vidalii, Wats. in Hook. Icon. 684. Azores only.

Isles 3. Flores (Capt. Vidal, R.N.); Watson cat. 113. Santa Maria and San Miguel; Hunt coll.

This remarkable and very distinct species apparently has not come under the notice of the German and French collectors. It was the only plant in my own collection which was not found by myself. Some few fragments of it were brought to me by Captain Vidal, picked on an insulated rock off the coast of Flores, northward (if I remember rightly) from the town of Santa Cruz. My own subsequent search for it on the main island of Flores was not successful; but it was difficult to examine well the steep sea-cliffs of this and other islands, constantly undermined by the wash of the ocean, and slipping down into it. The figure in Hooker's Icones, with the description, was made up from these fragments, and fails to convey a proper idea of the well-grown plant. Afterwards Mr. Hunt looked for the plant with better success, finding it very locally on the coasts of Santa Maria and San Miguel. Its introduction into English gardens was not from the originally discovered habitat of Flores, but from one or both of the other islands named. I have since cultivated it in my garden for a series of years. It succeeds in a frame without artificial heat, but must be completely protected from frost and from over-damp at the root. One of my first young plants, second year in its growth and not yet come to flowering, was carried to Kew. A distinguished Professor of Botany was walking in the gardens with Sir William Hooker. I challenged him to name the genus or order of my plant. He guessed it to belong to the order Proteaceæ! This anecdote is mentioned by way of showing that the *Campanula* is one of very peculiar growth, such as may warrant a full account and description of it here. Each group of Atlantic Isles (Azores, Madeira, Canaries, Cape Verdes) has its peculiar *Campanula,* those of Madeira and Canary affording technical characters for generic distinction, while that of the Azores is a true *Campanula,* though with the habit of a shrubby *Sempervivum.*

The first year from seed the stem of the young plant is erect and undivided, very few inches in height, with scattered obovate-spathulate leaves. Afterwards, either first or second year from seed, on attaining the height of three inches or thereabouts, the leaves become narrower and are collected into an evergreen terminal rosette, bearing some resemblance to the close-clustered leaves of a shrubby *Sempervivum.* The second or third year a whorl of 5 to 7 branches grows from the axils of the lower of these clustered leaves, and the main stem ceases to elongate, or almost ceases; the branches remain simple, and in their turn terminate in loose rosettes or clusters of leaves. The following year they elongate further, and again terminate in larger rosettes of leaves; a second whorl of branches growing from the axils of the upper leaves of the original single rosette close above the first whorl. The next year again the flower-stems are produced by a rapid growth of the apex of the rosettes which terminated the first branches. According to the vigour of the individual plant or branch, and the mass of leaves composing the rosette, the flower-branch becomes a simple raceme (unbranched panicle) of five or six flowers, with a few small leaf-like bracts, or it expands into a true branching panicle, one to two feet in length, with twenty or thirty flowers. The main stem and branches are at first succulent and brittle, gradually becoming hard and tough. The flower-branch and peduncles are extremely brittle during their early rapid growth from the rosettes of leaves, which apparently serve as a reservoir of nutriment for them. The whole plant is full of thick creamy juice, which is singularly viscid, and can be drawn out into long threads as it exudes on fracture. Perhaps this viscid juice would be worth the attention of chemists and doctors. The plant has been now traced up into a symmetrical shrub, having a short pedestal or main stem, 3 or 4 inches high, surmounted by rings of undivided branches, the lower ring ending in long panicles of large pendent flowers, the upper ring following the same course of development in the succeeding year or years. Usually, after three or four years, the primary rosette of leaves, terminating the short main stem, becomes exhausted and withers away; no more branches being produced from it; and the branches themselves die after flowering. Thus, stout and shrub-like as the plant is, its life is probably always short. I have not kept the same individual plant over five or six years. It renews itself by seed almost without attention; the very fine seed being abundant, and some of it pretty sure to germinate on the surface-soil of any flower-pots standing near, or anywhere else if left unburied and undisturbed, safe from frost and over-dampness. A detailed description of the grown plant is here subjoined to show the botanical or technical characters of its organs.

Stem erect, short, symmetrically branched round the top; branches alternate, approximated into annual rings of five or more, the first ring succeeded by second and third rings of younger branches. Leaves scattered, lanceolate-spathulate, serrate, closely crowded into rosettes at the summits of the main stem and branches, at first succulent, then passing to a leathery texture. Panicle terminating the branches either racemose with about half a dozen separately stalked drooping flowers, or the first peduncles branched and bearing several flowers on each; uppermost buds flowering the first. Calyx wide, succulent, without appendages. Corolla large, an inch or inch and a half long, corset-form, constricted in the middle, widening downwards and upwards, milk-white, often with a flesh-coloured or pink tint externally; divisions of the limb spreading or reflex, about one-fifth of the total length. Disk large, white, bordered by an orange ring. Stigmas three, linear-oblong. Capsule three-celled, half an inch wide across the disk or flattened top.

222. Campanula Erinus, Linn. South Europe.

Isles 3, or more. Miguel, Fayal, Flores. "Most of the islands;" Seubert flo. 232. Watson cat. 114. Hunt coll. Godman coll. Drouet cat. 264.

36. Vacciniaceæ.

223. Vaccinium cylindraceum, Sm. Azores only.

Isles 5. Miguel, Maria, Pico, Fayal, Flores. Seubert flo. 292. 29. Watson cat. 118. Hunt coll. Godman coll. Drouet cat. 367.

This is recorded under two or three names, and as so

many different species, in the Flora Azorica, in Drouet's Catalogue, and in the Prodromus of De Candolle. The *Vaccinium cylindraceum* of Smith, and *Vaccinium longiflorum* of Wickstroem, apparently mean just the same single species. The third specific name is *Maderense*, that of a closely allied species inhabiting Madeira, but not satisfactorily shown to be native also in the Azore Isles. All the laboured distinctions set forth in the Flora Azorica, in order to keep up two species out of *cylindraceum* and *longiflorum*, are mere verbiage. So far as there is any real difference between the characters there given, various intermediate gradations and cross characters occur among the plants. The drooping racemes and elongate flowers distinguish the Azore species from that of Madeira, in herbarium specimens; and the alleged more arborescent size or growth of the living shrub is said to afford an additional distinction for the Madeira species. At the time of writing this (June 1868) I have before me a weakly living shrub from San Miguel, just coming into flower, together with upwards of a score of dried specimens from various of the Azore Isles, selected to illustrate the variations of the species, also five specimens from Madeira. On contrasting the opposite extremes, the flowers do appear much dissimilar in length, and thus so far likewise in form, my longest Azore flowers being a full inch long, the smallest Madeira flowers less than half an inch. But an approximating gradation of size is readily traced, which leaves barely one-eighth of an inch between the longest flowers from Madeira and the shortest from the Azores. Whether a botanist seeking truth, rather than striving to make up book-species, would get over that one-eighth of an inch in a fuller series of examples from Madeira, I must leave to the decision of those who possess such a series. In the herbarium specimens above mentioned, I do not find any other invariable character to separate the Azore from the Madeira shrub, or any thing at all by which to divide the Azore examples into two species. The colour of the corolla is primarily a pale yellowish-green, becoming reddish or russet-purple where most exposed to light, some being entirely green, others nearly red externally, on the same single bush, even in the same raceme; or the two colours may be seen on the two contrary sides of a single corolla. To describe the flowers of the two pretended species of the Azores as being "rubra" and "pulchre rosea," as is done in the Flora, is simply to invent book-distinctions, neither of them true in nature. Precisely equivalent variations of tint are seen in the flowers of *Erica azorica*; and inconsistently enough these are treated only as varieties in the Flora Azorica; being, however, not even so much as true varieties, but only casual variations, arising from shade or exposure, and thus occasionally co-existent on the same single bush.

37. Ericaceæ.

224. Calluna vulgaris, Salisb. Europe. Britain. Isles 4, or more. Miguel, Pico, Fayal, Flores. "Almost all the islands;" Seubert flo. 288. Watson cat. 116. Hunt coll. Drouet cat. 363.

225. Menziesia polifolia, Sm. S.W. Europe. Ireland. Isles 4. Terceira, Pico, Fayal, Flores. Seubert flo. 289. 31. Watson cat. 117. Hunt coll. Godman coll. Drouet cat. 364.

My specimens from Fayal are quite dwarf, resembling the variety "*nana*" figured in Loddiges' Botanical Cabinet, 1907, the original habitat of which is there stated to be unknown.

226. Erica azorica, Hochst. Azores only?

Isles 5, or more. Miguel, Terceira, Pico, Fayal, Flores. "All the islands;" Seubert flo. 287. 30. Watson cat. 115. Hunt coll. Godman coll. Drouet cat. 362.

Erica scoparia var. *parviflora* of De Candolle's Prodromus, vol. vii. p. 692, according to Seubert, who also states that the true *Erica scoparia*, Linn., was found by Mr. Hochstetter on an islet near Villa Franca, in San Miguel. See *Corema alba*.

38. Oleaceæ.

227. Picconia excelsa, De Cand. Madeira. Canaries.

Isles 2, or 4. Miguel, Pico. Fayal?, Flores? Seubert flo. 241. 2. Watson cat. 119, "*Olea excelsa*." Hunt coll. Drouet cat. 270.

Mr. Hunt reports this truly native in San Miguel. In the smaller islands of Fayal and Flores I saw it only in places where it might likely have been planted.

39. Apocynaceæ.

228. Vinca media, Link. Portugal. Spain.

Isle 1. San Miguel; Hunt coll. Godman coll. Drouet cat. 280.

Published in my Supplementary List erroneously under the name of *Vinca major*, a very closely allied species.

40. Asclepiadaceæ.

229. Asclepias fruticosa, Linn. South Africa.

Isle 1. Fayal, certainly from gardens; Seubert flo. 242. 23. Watson cat. 120. Hunt coll. Drouet cat. 271.

Although thus recorded by most of the collectors among their plants observed in the Isles, it is there only as a casual escape from gardens. Now more usually known under the generic name of *Gomphocarpus*.

41. Gentianaceæ.

230. Erythræa Centaurium, Pers. Europe. Britain. Isles 5. Miguel, Maria, Terceira, Fayal, Flores. Seubert flo. 244. 4, "*latifolia*." Watson cat. 121. Hunt coll. Drouet cat. 274,—also probably 273 "*pulchella*" and 275 "*latifolia*."

It is possible that *E. pulchella* may occur in Santa Maria, as stated by Drouet's Catalogue; but it seems safer not to accept that habitat unless confirmed by additional authority. As to *E. latifolia* of Smith, correctly thus named, the only locality in which it has been certainly found is the neighbourhood of Liverpool, in England; and there it is now supposed to have become extinct. All the foreign examples thus labelled, and all the examples from other places in England which have come under my own observation, were simply broad-leaved forms of *Centaurium*. Smith's plant was very likely an irregular development only, not a true species.

231. Erythræa Massoni, Sweet. Azores only. Isles 5. Miguel, Terceira, Pico, Fayal, Flores. Seubert flo. 245. 12 & 13. Watson cat. 122. Hunt coll. Drouet cat. 276 & 278. Kew list of Godman coll.

Dr. Seubert, quoting Griesbach, makes this identical with *E. diffusa* of Woods. M. Drouet seizes on the two names to expand his own Catalogue, by giving them as two Azore species; pushing in the perfectly distinct "*maritima*" between them, and repeating this latter after them both, under name of "*lutea*." Certainly the *Massoni* is a widely varying species, as explained in my own former Catalogue. The fresh flowers are always white in the Azores, no pink or rose-coloured example having been observed by myself among the thousands which came under notice. The *diffusa* of Woods is not white-flowered; and the late Mr. P. B. Webb emphatically declared his dissent from a union of that and the *Massoni* as one species only.

232. Erythræa lutea, Roem. et Schult. S. Europe. Isles 1, or 4. Miguel; Hunt coll. Godman coll. "Santa Maria, Fayal, Pico;" Drouet cat. 277 "*maritima.*"

M. Drouet's confusion of the species in this genus leaves some degree of uncertainty whether his three habitats for "*maritima*" so called really belong to *lutea* or to something else. He includes San Miguel with the other three; but the habitat rests on authority far more satisfactory in my own eyes, namely, actual specimens sent by Mr. Hunt. At present, I can make out only three species of this genus in the Isles. These are subdivided into seven in Drouet's Catalogue. Seubert's Flora Azorica gives two of them, both under misnomers, I believe.

233. Exacum filiforme, Willd. Europe. England. Isles 2. Terceira; Seubert flo. 248. 8. Miguel; Hunt coll. (with slight uncertainty).

No example of this from the Isles is now in my own herbarium; but the name is marked in a list as that of a species sent to me from San Miguel by Mr. Hunt. The no. "8" in Seubert's Flora shows that specimens from Terceira were distributed in Hochstetter's collection.

42. Convolvulaceæ.

234. Convolvulus arvensis, Linn. Europe. Britain. Isles 2, or more. Miguel, Fayal. All the islands; Seu-

bert flo. 262. Watson cat. 123. Hunt coll. Drouet cat. 319. Kew list of Godman coll.

235. Convolvulus sepium, Linn. Europe. Britain. Isles 2. Fayal, Flores. Seubert flo. 263. 14. Watson cat. 124. Drouet cat. 347, but not marked as actually seen.

236. Batatas littoralis, Chois. South Europe. Isle 1. Fayal, in one spot, on the sandy shore by Porto Pym; Watson cat. 125, under synonym of *Convolvulus Imperati,* Vahl. Drouet cat. 320.

It is this plant which has given occasion to the erroneous record of "*Convolvules Soldanella*" in Drouet's 'Catalogue, where the *Soldanella* is entered as a native of Fayal, not seen by M. Drouet or his fellow travellers, and without the citation of any other authority. I presume the error to have arisen in the manner here set forth, as an example and a warning against too much trust being given to mere names in the Catalogue referred to. In a paper written before my own former Catalogue of Azore plants was made out, it was mentioned incidentally that I had seen on the sands about Porto Pym, in Fayal, " a *Convolvulus* much resembling *C. Soldanella,* but with white and larger flowers." Obviously, it was not thought to be truly *Soldanella,* or it would have been written of as a white-flowered variety of that species. I wrote as an Englishman, comparing the plant with its nearest affine or analogue in England, and mentioned a conspicuous difference, but not giving any ground for the inference of that being the only difference. It seems that Dr. Seubert introduced *Convolvulus Soldanella* into his own Flora solely on this intimation of a species much resembling it; but, at the same time, he faithfully repeated my statement of the Fayal species having white and larger flowers than *Soldanella.* In my own after printed Catalogue, I gave *Imperati* and not *Soldanella.* Nevertheless, M. Drouet puts the latter into his Catalogue, without a word of uncertainty, as a recorded Azore plant. So unreliable do authors become, where their aim is to make a long list, not to show truth simply as truth! To make the worst of the matter, he adds *Imperati* also as a second species, seen by himself in Fayal!

43. Solanaceæ.

237. Solanum nigrum, Linn. Europe. Britain. Isles 4. Miguel, Pico, Fayal, Flores. Seubert flo. 268. Watson cat. 145. Hunt coll. Kew list of Godman coll. Drouet cat. 324.

238. Solanum villosum, Lam. South Europe. Isle 1. Flores; Watson cat. 146. Apparently local; as not having been observed in other Isles, or by other collectors.

It came up true in the garden, from seeds brought to England; but seems to be hardly more than a variety of *S. nigrum.*

239. Solanum Pseudo-capsicum, Linn. Madeira. Isles 5. Miguel, Terceira, Pico, Fayal, Flores. Seubert flo. 269. 19. Watson cat. 144. Hunt coll. Godman coll. Drouet cat. 326.

Probably an introduced plant, but well established in various places by road sides.

240. Physalis pubescens, Linn. Madeira. Isles 4, or more. Miguel, Graciosa, Fayal, Flores. All the islands; Seubert flo. 267. 18. Watson cat. 147. Hunt coll. Godman coll. Drouet cat. 823. Introduced?

241. Hyoscyamus albus, Linn. var. South Europe.

Isles 6. Miguel, Maria, Graciosa, Pico, Fayal, Flores. Seubert flo. 266. 21. Watson cat. 148. Hunt coll. Godman coll. Drouet cat. 322.

This is the variety or subspecies *Canariensis*; which is so little different from states of the European *albus*, that I have preferred to take up the latter specific name.

242. Datura Stramonium, Linn. Europe?

Isles 2, or more. Wastes; Seubert flo. 265. "Fayal, San Miguel;" Drouet cat. 321, but not marked as actually seen. A casual alien?

44. Acanthaceæ.

243. Acanthus mollis, Linn. South Europe.

Isles 3. Miguel, Fayal, Flores. Watson cat. 162. Hunt coll. Not a native.

Introduced into my former Catalogue as being uncertainly an "Alien?" Mr. Hunt gives an affirmative answer to the query.

45. Scrophulariaceæ.

244. Verbascum virgatum, With. Europe. England.

Isles 3. Miguel; Hunt coll. Maria; Godman coll. Terceira; Drouet cat. 329.

Drouet's Catalogue enumerates also the *Verbascum Blattaria* as being found in Pico.

245. "Verbascum spurium, Koch!" Mid Europe?

Isle 1, or 2. San Jorge (Hartung); Drouet cat. 331. Apparently the only record; unless this is the same thing with an undetermined species sent from San Miguel by Mr. Hunt.

The sign of certainty affixed to the nominal authority

for this species is remarkable. By Koch, in the second edition of his Synopsis, *V. spurium* is made identical with *Thapso-Lychnitis*. But, since neither of the reputed parents has been reported as found in San Jorge, it is difficult to understand this alleged occurrence there of their supposed spurious progeny. Who would expect to find equine mules wild in an island where neither horses nor asses were seen? Still, in my own herbarium are two side branchlets from the flowering stem of a *Verbascum*, sent from San Miguel, which may be held allied both to *Thapsus* and to *Lychnitis*, but which is certainly neither of those species.

Verbascum Thapsus was enumerated in my former Catalogue as a plant of Fayal, probably an alien there. The non-native character is thus far confirmed by that single record remaining unsupported by a second.

246. Veronica Anagallis, Linn. Europe. Britain. Isles 3. Miguel, Fayal, Flores. Seubert flo. 278; but only on the quoted authority of "Watson." Watson cat. 160. Hunt coll. Godman coll. Drouet cat. 312.

247. Veronica Dabneyi, Hochst. Azores only. Isles 3. Miguel, Fayal, Corvo. Seubert flo. 279. Watson cat. 158. Hunt coll.

This species has been placed between *officinalis* and *Allionii* in the Prodromus of De Candolle, as if it had some close affinity with those species; and herbarium specimens seem to warrant the position. It is a diffusely branched rigid almost shrub-like plant, with thick evergreen leaves; and when seen alive, it bears no closer resemblance to *officinalis*, than the evergreen *Prunus Lauro-cerasus* of our gardens bears to the deciduous *Prunus Padus* of our woods. The flowers are always light pink colour, striped with lines of a deeper tint; and this colour has been kept unchanged through several successive descents in England. The plant endures our climate passably well; although the summers are too dry for it in Surrey, unless artificially aided by watering and shading. Severe winters are fatal to it, though it will live through our milder winters in the open ground. Like some others of the peculiarly Azore plants, its suitable climate here is found in a cold-frame, protected from frost in winter, and in a shaded somewhat damp situation in summer.

248. VERONICA OFFICINALIS, Linn. Europe. Britain. Isles 2. Pico, Fayal. Seubert flo. 277. 1. Watson cat. 157. Drouet cat. 343.

The supposed peculiarity of form in the capsule, remarked upon in the Flora Azorica, is frequently seen on English specimens of the species.

249. VERONICA SERPYLLIFOLIA, Linn. Europe. Britain. Isles 4. Miguel, Maria ("var. *pubescens*"), Fayal, Flores. Watson cat. 159. Hunt coll. Godman coll. Drouet cat. 344.

250. VERONICA ARVENSIS, Linn. Europe. Britain. Isles 3, or more. Miguel, Fayal, Flores. Everywhere in fields; Seubert flo. 276. Watson cat. 161. Hunt coll. Godman coll. Drouet cat. 345.

251. EUPHRASIA GRANDIFLORA, Hochst. Azores only. Isles 5. Terceira, Pico, Fayal, Flores, Corvo. Seubert flo. 280. 54. Watson cat. 151, under name of "*azorica.*" Drouet cat. 348. A much handsomer plant than the *E. officinalis*.

The Flora Azorica enumerates also the *Euphrasia officinalis* as observed "in pratis humidiusculis insularum plurium." If this record is really correct, it is curious that all the other collectors should have overlooked the plant, especially Mr. Hunt several years a resident in the largest island. Dr. Seubert was unable apparently to name any of the islands specially; and he gives no second no. to indicate its labelled distribution by Hochstetter. Under these circumstances, I take leave to disbelieve the record, while not warranted in contradicting it on the purely negative evidence.

252. BARTSIA TRIXAGO, Linn. South Europe. Isles 3. Miguel, Maria, Pico. Watson cat. 152. Hunt coll. Kew list of Godman coll. Drouet cat. 350.

253. BARTSIA VISCOSA, Linn. S.W. Europe. Britain. Isle 1. Terceira; Godman coll. 1865. An interesting addition to the flora of the Isles. Found also in the Canaries.

254. SCROPHULARIA SCORODONIA, L. S. Eur. England. Isles 2. Miguel, Terceira. Seubert flo. 271. 58. Hunt coll. Drouet cat. 333.

255. SCROPHULARIA BALBISII, Hornem. Eur. England. Isle 1. Flores; Watson cat. 133, under the synonym of *S. aquatica*. Apparently the only record, but certain.

256. ANTIRRHINUM ORONTIUM, Linn. Europe. England. Isles 5. Miguel, Maria, Terceira, Fayal, Flores. Seubert flo. 275. 57. Watson cat. 155. Hunt coll. Godman coll. Drouet cat. 340.

The Fayal specimens were nearly glabrous, with a purplish glaucous tint; corolla white, striped with purple, shorter than the calyx. The Flores specimens correspond with the English plant.

257. LINARIA ELATINE, Mill. Europe. England.
Isles 3, or more. Miguel, Terceira, Fayal. Seubert flo. 272, under name of "*Sieberi*." Watson cat. 156 in part. Hunt coll. Drouet cat. 355, "Terceira (Morelet)."

Dr. Seubert enumerates also *Linaria cirrhosa* 273 no. 56, as found near San Pedro, Terceira. This may be reserved for further confirmation. *Linaria dealbata* was given in my own former Catalogue and *Elatine* omitted; this latter certainly occurs; respecting the former I remain in considerable doubt.

258. LINARIA SPURIA, Mill. Europe. England.
Isles 4, or more. Miguel, Maria, Terceira, Fayal, Flores? Seubert flo. 274. Watson cat. 156 in part. Hunt coll. Godman coll. Drouet cat. 334.

259. SIBTHORPIA EUROPÆA, Linn. Europe. England.
Isles 4. Miguel, Pico, Fayal, Flores. Seubert flo. 275*. 27. Watson cat. 150. Hunt coll. Godman coll.

In the Prodromus of De Candolle the habitat of Azores is assigned to *S. africana*, not to *S. europæa*. Hitherto, I have been wholly unable to see or understand any species *africana* distinct from *europæa*. And independently of the Greek and other specimens labelled *africana* in my herbarium, I feel quite satisfied that the Azore plants are not even varieties of those found in England, Ireland, Jersey, and Portugal,—when a full series of examples is examined. The plant is well established in my garden from the Azore stock; having lived there through several years, on a shaded border set apart for hardy Ferns, growing so luxuriantly each summer as almost to smother some of the Ferns. I have not had the living plant of England actually growing with them; but all the pretended differences set forth in books break down when actual plants are looked at, whether living or dried, instead of attention being fixed on printed verbiage about them. For instance, the pretended diagnosis rests mainly on these alleged differences:—

Europæa . . . pedicellis brevissimis. Corolla, laciniis 2 flavidis, 3 roseis.

Africana . . . pedicellis petiolum æquantibus superantibusve. Corolla omnino flava.

What are the facts? In the living Azore plant, abundantly in flower now before me, the corolla is pale greenish-yellow, all the segments marked (unequally) with dull red. The petioles vary from a quarter of an inch or less to upwards of a full inch. The pedicels are as variable on the garden examples, and also in wild specimens; one of the latter from Fayal has the pedicels about a quarter-inch long, one from San Miguel has pedicels thrice that length. The European plants correspond. Bourgeau's no. 1974 from Portugal has some pedicels a full inch long, notwithstanding that "pedicellis brevissimis" are shown by the words that follow in the Prodromus to intend "usually less than a line long."

260. Digitalis purpurea, Linn. Europe. Britain.

Isle 1. Terceira (Morelet, Hartung); Drouet cat. 339. Local, or perhaps introduced; having escaped the notice of preceding collectors.

46. Verbenaceæ.

261. Verbena officinalis, Linn. Europe. Britain. Isles 4, or more. Miguel, Pico, Fayal, Flores. Seubert flo. 255. Watson cat. 163. Hunt coll. Godman coll. Drouet cat. 307.

47. Lamiaceæ.

262. Mentha rotundifolia, Linn. Europe. Britain. Isles 3, or more. Miguel, Fayal, Flores. Seubert flo. 248. Watson cat. 129. Hunt coll. Drouet cat. 282.

263. Mentha sativa, Linn.? Europe. Britain. Isles 3, or more. Miguel, Fayal, Flores. Seubert flo. 247? Watson cat. 131. Hunt coll.

The *Mints* remain still one of my difficulties in the flora of the Isles. Seubert enumerates two only; placing them under the names of *rotundifolia* and *viridis*, and without specifying any island by name as the habitat of either. Probably his materials were insufficient, equally with those on which the names of *sativa* and *aquatica* were given in my own former Catalogue. Mr. Hunt sent many specimens from San Miguel; and for these I still doubtingly continue the names before used for specimens from Fayal and Flores; omitting *viridis* as too uncertain. Dr. Mackay sent from Flores one non-flowering specimen apparently of *viridis*; but I suspect this not to have been a truly wild example. Dr. Seubert gives the habitat of his *Mentha viridis* "in dumetis humidis plurium insularum," a description perhaps transcribed from some book on European botany with the addition of the two last words for difference of place. Drouet's Catalogue adds still another species, *piperita*, as found by Hartung in Santa Maria;—

not likely, unless there as an introduced plant become wild.

264. Mentha aquatica, Linn.? Europe. Britain. Isles 3. Miguel, Fayal, Flores. Watson cat. 132. Hunt coll. Drouet cat. 285.

None of my specimens correspond precisely with those in my herbarium from British or European localities. But they come nearer to this species than to any thing else known to me.

265. Mentha Pulegium, Linn., et var. Eur. England. Isles 6: Miguel, Maria, Pico, Fayal, Flores, Corvo. Watson cat. 133 and 134. Hunt coll. Kew list of Godman coll. Drouet cat. 287.

It is remarkable that so frequent a plant of the Isles should not have been enumerated in Seubert's Flora. It occurs in two forms, the common trailing plant as seen in Britain, but chiefly the larger, more upright, and very downy form of the Mediterranean countries.

266. Lycopus europæus, Linn. Europe. Britain. Isles 2. San Miguel; Hunt coll. Terceira (Morelet); Drouet cat. 288.

267. Thymus angustifolius, Pers. South Europe. Isles 6. Miguel, Maria, Terceira, Pico, Fayal, Flores. Seubert flo. 251. 49. Watson cat. 136. Hunt coll. Godman coll. Drouet cat. 294.

This is the *Thymus micans* of the Flora Azorica and of the two Catalogues cited; a subspecies or segregate, the name of which is reduced to a synonym in the Prodromus of De Candolle.

268. Origanum virens, Link. South Europe.

Isles 4, or 6. Miguel, Maria, Terceira, Flores. Pico? Fayal? Watson cat. 139. Hunt coll. Godman coll. Drouet cat. 294.

This is omitted from the Flora Azorica. Instead, a plant found in Pico is enumerated there under name of *O. creticum*. In Drouet's Catalogue we have the further additions of *vulgare* alleged to be found in Fayal, and of *Marjorana* for Pico. The latter species is cultivated in the Isles, and possibly the two other names *vulgare* and *creticum* may intend the plant here named *virens*.

269. Calamintha officinalis, Moench., var.
Europe. Britain.

Isles 6. Miguel, Maria, Terceira, Graciosa, Fayal, Flores. Seubert flo. 252. 51. Watson cat. 135. Hunt coll. Drouet cat. 296?

There is some difficulty with this plant, arising out of the diversities in naming the European varieties or subspecies. It is not the *Calamintha sylvatica* of Bromfield, which is labelled as true *officinalis* by some good European botanists. It corresponds well enough with the *Calamintha ascendens* of Jordan, and likely is the plant recorded as the *officinalis* in the Prodromus of Candolle. There are two states of it in the Isles, differing in the density of the pubescence; one of them looking more hoary than the other. In the Catalogue of Drouet it is given as *Melissa Calamintha* var. *villosissima,* and apparently also as *Calamintha nepeta* var. *rotundifolia.*

270. Calamintha Clinopodium, Benth. Eur. Britain.

Isles 3. Miguel, Maria, Flores. Watson cat. 137.

Hunt coll. Kew list of Godman coll. Drouet cat. 295. *Clinopodium vulgare,* Linn.

271. Ballota nigra, Linn. Europe. Britain. Isle 1. San Miguel; Hunt coll. Mr. Hunt's specimens belong to the common European form *fœtida,* not to *ruderalis.*

272. Lamium purpureum, Linn. Europe. Britain. Isle 1. San Miguel; Hunt coll. Overlooked or disregarded by other collectors.

273. Lamium amplexicaule, Linn. Europe. Britain. Isle 1. San Miguel; Hunt coll. Found also by Mr. Godman in the same island.

274. Stachys arvensis, Linn. Europe. Britain. Isles 4. Miguel, Terceira, Fayal, Flores. Seubert flo. 254. Watson cat. 140. Hunt coll. Drouet cat. 304.

275. Nepeta Glechoma, Benth. Europe. Britain. Isle 1. San Miguel; Hunt coll. The *Glechoma hederacea* of Linnæus and of most British and European writers.

276. Marrubium vulgare, Linn. Europe. Britain. Isles 4. San Miguel; Hunt coll. Terceira; Godman coll. Santa Maria and Graciosa; Drouet cat. 305.

277. Prunella vulgaris, Linn. Europe. Britain. Isles 5. Miguel, Maria, Pico, Fayal, Flores. Seubert flo. 253. 53. Watson cat. 141. Hunt coll. Godman coll. Drouet cat. 303.

48. Boraginaceæ.

278. Myosotis azorica, Wats. Azores only.

Isles 2. Flores, Corvo. Watson cat. 128. Godman coll. Probably quite local; for the other collectors would not have disregarded it if seen in flower.

Some variations have occurred in the size and tint of the corolla; but otherwise it remains permanently distinct under culture in England. A very difficult plant to cultivate; not bearing the dry-heat of summer or the frost-cold of winter in Surrey, unless protected from both. When successfully grown it becomes a beautiful plant; usually it is seen only as a sickly deformity, damaged by insects or slugs, and struggling against an unsuitable climate. The rich deep blue or indigo colour of the flowers, with their abundance on vigorous examples, makes it quite a different plant from the pale-flowered species of Europe. Its place of growth in the Isles was on damp rocks facing from the sun, or alongside of mountain-streams. *M. azorica* and *M. maritima* are in some sort intermediates in habit between the perennial marsh species and the annual dry-ground species of Europe; both of them being quite distinct from the European species. The Latin of my diagnosis in the Botanical Magazine, where this plant is figured, has been strangely miscopied or misprinted after leaving my own hands.

279. Myosotis maritima, Hochst. Azores only.

Isles 2. Pico, Fayal. Seubert flo. 261. 9. Watson cat. 127. Drouet cat. 315.

Some details about this species, as grown in England, were given in my Supplementary List and Notes on Azore Plants, published in the London Journal of Botany, vol. vi. p. 388.

280. Myosotis arvensis, Anglorum. Europe. Britain. Isle 1. San Miguel; Hunt coll. The *M. intermedia* of various European authors.

281. Myosotis stricta, Link. Europe. Isle 1, or more. "Hinc inde in apricis;" Seubert flo. 259. Is it in the Kew herbarium, collected in Fayal by Mr. Godman? "Not there;" Rev. W. W. Newbould, msc.

282. Myosotis versicolor, Pers. Europe. Britain. Isles 2. San Miguel; Seubert flo. 260. Santa Maria; Drouet cat. 314. Mr. Hunt likewise sent it from the island of San Miguel.

283. Cynoglossum pictum, Ait. South Europe. Isles 4. Miguel, Maria, Terceira, Flores. Hunt coll. Godman coll. Drouet cat. 309.

284. Heliotropium europæum, Linn. South Europe. Isles 3. Terceira, Graciosa, Pico. Seubert flo. 256. 10. Watson cat. 126. Godman coll. Drouet cat. 308.

285. Echium violaceum, Linn. South Europe. Isles 3. Miguel, Maria, Terceira. Seubert flo. 257. Hunt coll. Godman coll. Drouet cat. 310.

The Flora Azorica enumerates also *Echium vulgare*, but without record of any habitat, except the vague words "hinc inde"; words, it may be feared, too frequently introduced at random into the work. Drouet's Catalogue assigns it to San Miguel.

49. Myrsinaceæ.

286. Myrsine africana, Linn., var. retusa. Africa. Isles 5, or more. Miguel, Pico, Fayal, Flores, Corvo. "All the islands;" Seubert flo. 285. 122. Watson cat.

169. Hunt coll. Godman coll. Drouet cat. 360. *Myrsine retusa* (Aiton) of the two Catalogues. *Myrsine africana* var. *retusa* of De Cand. prodr. 8, 93, and Seubert flo.

50. Primulaceæ.

287. Lysimachia azorica, Hornem. Azores only. Isles 5. Miguel, Maria, Terceira, Fayal, Flores. Seubert flo. 283. 15. Watson cat. 164. Hunt coll. Drouet cat. 352 and 353. Kew list of Godman coll.

This is nearly allied to *L. nemorum* of Europe, and not very readily to be distinguished in description from that species; partly because the plant of the Isles is itself variable, and more so than the European species. By Seubert this is advisedly made identical with *nemorum*, Linn., as if not distinguishable even as a variety. After twenty years' culture of it in England, I hold it specifically and permanently distinct from our British and European *nemorum*. There are two principal forms in the Isles. That which may be deemed the type or prevailing form differs from *nemorum* in the following characters:—Stems ascending from a procumbent base, scarcely at all rooting at the joints; leaves of a lighter green colour, narrower and more blunt, closely approximated on the stem so as to leave shorter internodes, and towards the ends of the branches so close as to be reflexedly imbricated; flowers larger, and more star-like by the longer, narrower, and more separated divisions of the corolla; sepals wider, narrowly elliptic rather than subulate. The other form or variety approaches nearer to *nemorum* by its more trailing stems slightly rooting at some of the joints, and by the sepals being sometimes narrower and more inclining to subulate. This variety, however, recedes from *nemorum* in other respects; especially so by its star-like flowers and its firm imbricated pale-green leaves. Subvarieties of each form occur with leaves so narrow as to be lanceolate-elliptic, quite unlike the broadly ovate or subcordate leaves of *nemorum*. Although not thoroughly hardy in our climate, it survives mild winters in the open ground, and remains green through the winter if protected in a cold frame. True *L. nemorum* is not found in the Isles.

288. ANAGALLIS ARVENSIS, Linn. Europe. Britain. Isles 6. Miguel, Maria, Graciosa, Pico, Fayal, Flores. Seubert flo. 284. Watson cat. 165. Hunt coll. Drouet cat. 354 and 356. Kew list of Godman coll.

It is an illustration of M. Drouet's peculiar tendency to lengthen his list by nominal species, that he repeats this familiar European plant twice over, first under the Linnean name of *arvensis* as actually seen, and then under the synonym of *phœnicea* as also seen; *A. cærulea* being interposed between those two, as a species not seen by himself or companion. Some botanists certainly make the old name *arvensis* include or apply to both the other two, the blue and the red; but what can be intended by *arvensis* additional to these two?

289. ANAGALLIS CÆRULEA, All. Europe. England? Isles 2, or more. Miguel, Fayal, &c. United with *A. arvensis* in Seubert flo. 284. Watson cat. 166. Hunt coll.

This blue-flowered plant is less boreal than the red-flowered one on the Continent as well as in Britain; but there is little character to distinguish them, excepting the obvious one of colour.

290. ANAGALLIS TENELLA, Linn. Europe. Britain. Isles 2. Santa Maria, Pico; Drouet cat. 357. Not recorded by other collectors.

This plant cannot be declared a very unlikely one to occur in the Isles. It is not easy to point out any other plant likely to be mistaken for the *tenella*; while its small size and creeping manner of growth would render it not unlikely to be overlooked. Still it must remain desirable to have the two habitats verified by some other observer more specially giving his attention to botany.

291. Centunculus minimus, Linn. Europe. Britain. Isles 2. Flores, Corvo; Watson cat. 167. Not otherwise recorded. It is scarcely possible that the *Centunculus* could have been mistaken for the preceding; and the habitats are different.

292. Samolus Valerandi, Linn. Europe. Britain. Isle 1. Flores; Watson cat. 168. This plant, so widely distributed over the earth, would seem to be quite local in the Isles, as seen by only a single collector among the eight or ten.

51. Plantaginaceæ.

293. Plantago major, Linn. Europe. Britain. Isles 4, or more. Miguel, Graciosa, Fayal, Flores. "All the islands;" Seubert flo. 189. Watson cat. 206. Hunt coll. Godman coll. Drouet cat. 368.

In Flora Azorica it is stated that *Plantago media* occurs with this, "cum præcedente." By implication thus, it should be found in "all the islands." But no other collector appears to have seen the *media*; and there is no indication in the Flora that any specimens of it were distributed by Hochstetter. Under these circumstances, I take leave to hold the fact of its occurrence in the Isles at all as one needing verification.

294. Plantago lanceolata, Linn. var. Europe. Britain. Isles 4. Miguel, Terceira, Fayal, Flores. Seubert flo. 191 and 192 "*azorica.*" Watson cat. 207. Hunt coll. Godman coll. Drouet cat. 272.

Plantago azorica, of Seubert's Flora, is likely the prevailing form in the Isles. Having raised it in England, in successive years, from seeds on my specimens from Fayal, I find that the characters described by Seubert for a diagnosis are inconstant; in part certainly so, if not wholly so. *Plantago Lagopus* is also enumerated in the Flora, with only the vague intimation that it occurs with the rest, "cum prioribus;" no island being named for it, and apparently no specimens of it having been distributed, since Dr. Seubert fails to cite the number of any label for it.

295. Plantago Coronopus, Linn. Europe. Britain. Isles 4. Miguel, Terceira, Fayal, Flores. Seubert flo. 188. Watson cat. 208. Hunt coll. Godman coll. Drouet cat. 370.

296. Plantago Serraria, Linn. South Europe. Isle 1. San Miguel; Hunt coll. Apparently not observed by any other collector; but it may have been passed by as a large state or variety of *Coronopus*; and, indeed, I was so inclined to label it, until assured by Dr. Lemann that it is the *Serraria*.

297. Littorella lacustris, Linn. Europe. Britain. Isle 1. Corvo; Watson cat. 109. Perhaps overlooked elsewhere and by other collectors.

This is another of the plants which tend to establish a more close affinity between the Azorean and European floras, than is seen in those of Madeira or the Canaries.

52. Plumbaginaceæ.

298. Statice Limonium, Linn. var. Europe? Britain? Isles 3. Miguel, Maria, Terceira. Hunt coll. Godman coll. Drouet cat. 376 and 377.

Possibly two species are included here, neither of which is quite certainly the same with the *Limonium* of Britain. No species of *Statice* was recorded in the Flora Azorica. In Drouet's Catalogue the two names of *Limonium* and *serotina* represent two species.

53. Phytolaccaceæ.

299. Phytolacca decandra, Linn. South Europe. Isles 4. Miguel, Graciosa, Fayal, Flores. Seubert flo. 177. Watson cat. 224. Hunt coll. Godman coll. Drouet cat. 379.

Well established in Flores, by the beds of streams dry in summer; seen also in Fayal, where it appeared to be less clearly spontaneous. Doubtless an introduced plant to the Isles.

54. Salsolaceæ.

300. Beta maritima, Linn. Europe. Britain. Isles 3. Miguel, Maria, Fayal. Watson cat. 215. Hunt coll. Drouet cat. 388.

301. Chenopodium murale, Linn. Europe. Britain. Isles 6. Miguel, Maria, Terceira, Pico, Fayal, Corvo. Seubert flo. 180. Watson cat. 212. Hunt coll. Godman coll. Drouet cat. 385. Terceira; Baron do Castello de Paiva, in Kew herbarium.

302. Chenopodium ambrosioides, Linn. S. Europe. Isles 4, or more. Miguel, Terceira, Fayal, Flores. Fre-

quent; Seubert flo. 178. Watson cat. 211. Hunt coll. Godman coll. Drouet cat. 384. Introduced to Europe, and likely so to the Isles.

303. CHENOPODIUM RUBRUM, Linn. Europe. Britain. Isle 1, or more. Several of the islands, infrequent; Seubert flo. 179. Santa Maria; Drouet cat. 386. Not found by the three or four English collectors; and we English may thus hold it desirable to get a confirmation.

304. ATRIPLEX BABINGTONII, Woods? var. Eur.? Britain. Isles 3. Miguel, Flores, Corvo. Watson cat. 214. Hunt coll. Dr. Mackay! Drouet cat. 390?

This was enumerated in my former Catalogue under the name of *patula*; intending thereby the *patula* of Smith and of most English writers until quite recently; but which was more lately ascertained to be synonymous with *hastata* of the Continental botanists, who apply the name *patula* to the *angustifolia* of Smith. Having received better examples of it from Mr. Hunt, and having raised plants of it in the garden from seeds on Mr. Hunt's specimens sent from San Miguel, I am now disposed to assign the Azore plant to *Babingtonii*, although not with entire confidence. No species of *Atriplex* was given in the Flora Azorica. By Drouet's Catalogue, the *patula* is located in Flores and Corvo; the name not marked as that of a plant actually seen. The *Atriplex* of San Miguel is there enumerated as a variety of *portulacoides*, but "different from the type by its hastate leaves." Apparently Drouet intends the species presently under consideration, the leaves of which are hastate entire and the calyx valves also entire and smooth; so far as yet appears, it is the only *Atriplex* in the Isles.

305. Salsola Kali, Linn. Europe. Britain. Isles 2. Miguel, Fayal. Seubert flo. 181. Watson cat. 213. Hunt coll.

55. Amaranthaceæ.

306. Amaranthus Blitum, Linn. Europe. England. Isles 5. Miguel, Terceira, Fayal, Flores, Corvo. Watson cat. 210. Hunt coll. Godman coll. Drouet cat. 380.

307. Amaranthus chlorostachys, Willd. S. Europe. Isle 1. Near Horta, in Fayal; the single specimen alluded to in my former Catalogue, as being perhaps *A. strictus.* A specimen from Madeira, labelled "*strictus*" by the late Dr. Lemann, appears to be the same species.

308. Euxolus deflexus, Rafin. South Europe. Isles 3. Miguel; Hunt coll. Terceira and Fayal; Drouet cat. 381, under name of *Amarantus prostratus.*

309. Achyranthes argentea, Lam. South Europe. Isle 1. San Miguel; Hunt coll. Not in Fayal, where it is erroneously located by Drouet's Catalogue, and seemingly on my testimony.

310. Alternanthera Achyrantha, Brown. Spain. Isles 2. Miguel; Hunt coll. Terceira; Drouet cat. 383. Erroneously entered in my Supplementary list under name of *A. polygonoides.*

311. Illecebrum verticillatum, Linn. Eur. England. Isle 1, or more. "Hab. in inundatis editioribus"; Seubert flo. 329. Terceira; Godman coll. The name also occurs in a manuscript list of Guthnick's published collection.

56. Polygonaceæ.

312. Polygonum serrulatum, Lagasca. S. Europe.

Isles 3, or 4. Miguel, Flores, Corvo. Terceira? Watson cat. 223 and Supplementary Notes. Hunt coll.

I suppose this to be the "*Polygonum Persicaria*" of the Flora Azorica no. 183, without any island specially named as the habitat for it; also, the "*Polygonum dubium* Stein" of Drouet's Catalogue, located in Terceira. Still, the *Persicaria* seems not unlikely to occur in the Isles as an imported weed. Some of Mr. Hunt's specimens were distributed in England labelled with the temporary name of "*azoricum*" for distinction.

313. Polygonum maritimum, Linn. South Europe.

Isles 3, or more. Miguel, Pico, Fayal. All the islands; Seubert flo. 182. Watson cat. 222. Hunt coll. Godman coll. Drouet cat. Pico; Baron do Castello de Paiva, in Kew herbarium.

314. Polygonum aviculare, Linn. Europe. Britain.

Isles 5. Watson cat. 221. Hunt coll. Godman coll. Drouet cat. 394.

315. Rumex "aquaticus, Linn." Europe. Britain.

Isles 2. Fayal; Watson cat. 216. San Miguel; Hunt coll. "*Rumex Caldeirarum*" on the labels with some of Mr. Hunt's specimens; without specific name in Watson's Catalogue cited.

Dr. Meisner places this fine Dock, and also the *Hippolapathum* of Fries, under the Linnean *Rumex aquaticus*. The three certainly appear to be closely allied; though I cannot feel quite satisfied of the Azore species being identical with the boreal *Rumex aquaticus*.

316. Rumex crispus, Linn. Europe. Britain.

Isles 2. Miguel; Hunt coll. Corvo; Watson cat. 218. No record for the intermediate islands.

The Flora Azorica doubtingly enumerates *Rumex strictus* (or *dentatus*, an eastern species) as being found in Flores. Is this either the present or preceding one?

317. Rumex conglomeratus, Murr. Europe. Britain.

Isles 4. Miguel, Maria, Fayal, Flores. Watson cat. 217. Hunt coll. Drouet cat. 396 and 397.

Under the synonym of *Rumex acutus* of Smith in my former Catalogue, which doubtless misled Drouet into the error of making this plant into two species under the two names.

318. Rumex pulcher, Linn. Europe. England.

Isles 2. Miguel, Fayal. Watson cat. Hunt coll. Godman coll.

319. Rumex bucephalophorus, Linn. South Europe.

Isles 3, or more. Miguel, Pico, Terceira. All the islands; Seubert flo. 184. Hunt coll. Godman coll. Drouet cat. 399. Pico; Baron do Castello de Paiva, in Kew herbarium.

Excepting the uncertain *Rumex strictus* above mentioned, this present one is the only species of *Rumex* enumerated in the Flora Azorica. Stated to occur in all the islands, but was not observed anywhere by myself, while I did see the other five species which do not find mention in Flora Azorica.

320. Rumex Acetosella, Linn. Europe. Britain.

Isles 5. Miguel, Maria, Terceira, Fayal, Flores. Watson cat. 220. Hunt coll. Godman coll. Drouet cat. 400. Flores; Baron do Castello de Paiva, in Kew herbarium.

57. Thymelæaceæ.

321. Daphne Laureola, Linn. Europe. Britain.

Isle 1. Pico, at a considerable elevation. Seubert flo. 187. 28. Watson cat. 227. Drouet cat. 403.

A slight variety, with more spreading branches than seen in England; the variation possibly to be attributed to its place of growth, on the open declivity of a mountain, not sheltered in woods. M. Drouet was told that formerly it occurred likewise in San Miguel.

58. Lauraceæ.

322. Laurus canariensis, Webb.—" nec Willd."
Madeira. Canaries.

Isles 4, or more. Miguel, Pico, Fayal, Flores. All the islands; Seubert flo. 186. 33. Watson cat. 226. Hunt coll. Godman coll. Drouet cat. 404.

Under the name of *Persea azorica* in the Flora, and so repeated in the two Catalogues cited. The leaves vary much in shape and obtuseness; those of the female plant being more elliptic, those of the male plant more ovate, but variable in both.

323. Persea indica, Sprengel. Madeira. Canaries.

Isles 5. Miguel, Maria, Terceira, Fayal, Flores. Watson cat. 225. Hunt coll. Drouet cat. 405.

The true nativity of this handsome tree in Fayal, Flores, where it was seen by myself, appeared rather questionable. In other islands, according to M. Drouet, there exist "entire woods of it." Reputed native in Madeira and Canaries.

According to Drouet's Catalogue, *Oreodaphne fœtens* was seen in Terceira (bois des Garridas) by Hartung.

59. Empetraceæ.

324. Corema alba, Don. Portugal.
Isles 2, or 3. Pico; Watson cat. 228. Fayal; Drouet cat. 407. Miguel; Baron do Castello de Paiva, in Kew herbarium; but the specimen was labelled *Erica scoparia*, —which may be only a simple misnomer, leaving the habitat correct,—or, it may be a crossing of labels and specimens.

60. Euphorbiaceæ.

325. Euphorbia mellifera, Ait. var. Madeira. Canaries.
Isles 5. Miguel, Pico, Fayal, Flores, Corvo. Seubert flo. 172. Watson cat. 230 "Stygiana." Hunt coll. Godman coll. Drouet cat. 418.

As remarked in my former Catalogue, the specimens collected by myself in the more westerly islands, and subsequently others by Mr. Hunt in San Miguel, all differed conspicuously from the only two Madeira specimens of *mellifera* in my herbarium. The leaves of the Azore shrub are larger and thicker, dark or purple-glaucous, with stronger midrib ending in a short wide mucro. It would seem, however, from Mr. P. B. Webb's report, that a similar variety exists in the Canaries, and there attains an arborescent stature. The name in my Catalogue was misprinted *Stygiana*, which should have been *Styx-iana*, as it was intended to commemorate the steam-vessel *Styx*, from which I landed to botanize. By way of classical improvement Mr. Webb changed the name to "Stygia;"—quite departing from its intended significance. I might as well classicize the specific name *Webbiana* into *textilis*.

326. Euphorbia Lathyris, Linn. Europe. (England.)
Isles 3. San Miguel, near a village; Hunt coll. Pico,

Fayal; Seubert flo. 170. Likely a casual escape from culture.

327. EUPHORBIA PEPLIS, Linn. Europe. England.
Isles 3. Miguel, Pico, Fayal. Seubert flo. 169. 118. Watson cat. 234. Hunt coll. Drouet cat. 440. Kew list of Godman coll.

328. EUPHORBIA EXIGUA, Linn. Europe. Britain.
Isles 2. San Miguel; Hunt coll. Fayal; Watson cat. 232;—also, Hartung in Drouet cat. 413. Kew list of Godman coll.

329. EUPHORBIA PEPLUS, Linn. Europe. Britain.
Isles 3. San Miguel, Terceira, Flores. Watson cat. 233. Hunt coll. Godman coll. Drouet cat. 414.

330. EUPHORBIA AZORICA, Hochst. Azores only.
Isles 4, or more. Miguel, Pico, Fayal, Flores. Many of the islands; Seubert flo. 171. 119. Watson cat. 231. Hunt coll. Godman coll. Drouet cat. 417.

A somewhat variable species; different from any other known to me in this numerous and difficult genus, almost an order in itself. In De Candolle's Prodromus it is given by Boissier as a variety of *E. pinea,* Linn. Drouet's Catalogue enumerates also the following as Azoric:—

Esula, Linn.? 412. San Jorge (Hartung).

Gerardiana, Jacq. 411. Fayal.

Portlandica, Linn. 415. Terceira (Morelet).

No other collector appears to have found these three.

331. MERCURIALIS ANNUA, Linn. Europe. England.
Isles 2. Miguel, Fayal, Watson cat. 235. Hunt coll. Godman coll. Drouet cat. 419.

61. Urticaceæ.

332. Urtica membranacea, Poir. South Europe. Isles 4. Miguel, Terceira, Pico, Fayal. Seubert flo. 174 and 175. 120. Watson cat. 238. Hunt coll. Godman coll. Drouet cat. 420, 421, 422.

This is recorded in Flora Azorica under the specific names of *azorica* and *Lowei*. In Drouet's Catalogue those two names are repeated with the addition of *membranacea*; thus making out three nominal species for the Isles. I can see only a single variable species in my herbarium specimens from San Miguel and Fayal; some of which were referred to *neglecta* of Gussone, by the late Mr. Webb and Dr. Alexander. The examples destitute of the flattened male spikes look very like our British *Urtica urens*; and these perhaps may be what was intended by name of Gussone's species.

333. Parietaria officinalis, Linn. Europe. Britain. Isles 4. Miguel, Maria, Fayal, Flores. Seubert flo. 176. Watson cat. 237. Hunt coll. Godman coll. Drouet cat. 427.

334. Parietaria lusitanica, Linn. South Europe. Isles 2. Miguel; Hunt coll. Pico; Watson cat. 236. Apparently not observed by other collectors.

62. Ceratophyllaceæ.

335. Ceratophyllum demersum? Europe. Britain. Isle 1. Flores; Watson cat. 85. A specimen in leaf only, and the specific name thus being uncertain.

63. Amentiferæ.

336. Myrica Faya, Aiton. Madeira. Isles 4, or more. Miguel, Pico, Fayal, Flores. All the

islands; Seubert flo. 167. Watson cat. 241. Hunt coll. Godman coll. Drouet cat. 428.

In Flora Azorica this is described as a tall tree, "arbor procera." According to my own recollections it would be more truly described as a dense bush than a tall tree. But, like our native *Ilex Aquifolium,* it may occur under both conditions of growth.

Salix fragilis and *Populus nigra,* enumerated in my former Catalogue, are omitted from the present one, as being only planted trees in the Isles. I presume the same to be the case with *Ulmus campestris,* included in Drouet's Catalogue.

64. Coniferæ.

337. Juniperus brevifolia, Hochst. Spain? Madeira? Isles 4, or more. Miguel, Pico, Fayal, Flores. Seubert flo. 163. 124. Watson cat. 242. Hunt coll. Godman coll. Drouet cat. 429.

The affinity of the Azore Juniper is scarcely yet decided, although it has been much thought about. In Flora Azorica it stands as a variety *brevifolia* of the Linnean *Oxycedrus.* They bear much the same sort of resemblance to each other, as we see in our native Junipers, the *communis* and *nana.* The Azore Juniper is a very compact bush or small tree, ordinarily with upright stem, but procumbent on exposed hill-summits, where I have walked on the spreading and interlacing branches for several yards together, without touching the rocky surface of the ground underneath; stepping from branch to branch and from bush to bush. The leaves are wide and blunt, in comparison with those of the South European *Oxycedrus,* and only half of their length. But the transition is slight from the Azore form to the single Madeira example in my herbarium, received from Mr. Johnston as *Oxycedrus.* And the passage from this again to the Canarian examples is also slight. While in turn those from Canaries come nearer to the Hispanian, especially to my specimen no. 62 of Welwitsch's Flora Lusitanica. The Juniper of the Azore Isles seems a wider divergence from the European *Oxycedrus*, than are the Junipers of Madeira or the Canaries; but I write this without having had the opportunity to see the living trees from these latter islands, and have examined only a few herbarium specimens from them. Professor Parlatore allows it to stand as a distinct species in De Candolle's Prodromus, under name of "*J. brevifolia* (Antoine Cupress. Gattung.)."

As yet I have not obtained any confirmation of the report that *Taxus baccata* occurs wild on the mountains of these Isles; and I believe it may be truly asserted that no species of *Pinus* is wild there, though *Pinea* is included in Flora Azorica, as uncertainly so.

65. ORCHIDACEÆ.

338. SERAPIAS CORDIGERA, Linn. South Europe.
Isles 5. Miguel, Maria, Terceira, Pico, Fayal. Seubert flo. 157. 115. Watson cat. 243. Hunt coll. Godman coll. Drouet cat. 444.

339. HABENARIA MICRANTHA, Hochst. Azores only.
Isles 5. Miguel, Maria, Pico, Fayal, Flores. Seubert flo. 155. Watson cat. 244. Hunt coll. Godman coll. Drouet cat. 442.

340. HABENARIA LONGEBRACTEATA, Hochst. Azores only.
Isles 3. Miguel, Maria, Flores. Grassy places; Seubert flo. 156. 114. Flores; Watson cat. 244 in part. Miguel, Maria; Drouet cat. 443.

66. Iridaceæ.

341. Iris fœtidissima, Linn. Europe. England.

Isles 3. Miguel, Maria; Drouet cat. 445. Fayal; Baron do Castello de Paiva in Kew Herbarium?

The *Iris* mentioned in my former Catalogue, as seen near Largens in Flores, doubtfully indigenous, was probably *germanica*; and may be omitted from the present list as very likely an introduced species. The same course also may be adopted with the *Gladiolus* seen near Flamingos in Fayal, which was most likely an escape from garden culture.

342. Trichonema Columnæ, Reichenb. Eur. England.

Isle 1. San Miguel, at 1000–1500 feet; Hunt coll. An early flowering plant, very likely to be overlooked by summer visitants in the Isles.

67. Amaryllidaceæ.

343. Amaryllis Belladonna, Linn. South Africa.

Isles 3. Miguel, Terceira, Fayal. Seubert flo. page 25. Watson cat. 247. Drouet cat. 449.

This ornamental plant is cultivated in gardens, and it may be found occasionally by road-sides where garden refuse is thrown. On sandy ground along the shore in Fayal the bulbs are plentiful, and may be seen in summer lying about on the dried sand. The *Agave americana* is enumerated in Drouet's Catalogue as found in Santa Maria; doubtless an introduction from America, either direct or through Portugal. The *Narcissus* observed on Monte Carneiro in Fayal, was also very likely an introduction through gardens. It corresponds well with the specimens labelled *Narcissus stellatus* in Dr. Welwitsch's Flora Lusitanica, nos. 226 and 972.

68. Liliaceæ.

344. Allium Ampeloprasum, Linn. Europe. (England.) Isles 2. San Miguel; Hunt coll. I found some few plants of this species, as supposed (possibly, *Babingtonii* or *Porrum*) in an early stage on the coast-cliffs near a village in Fayal. On attempting to dry examples in my small ship-cabin, the scent proved so disagreeable to myself and neighbours as to compel me to throw them overboard. I did not learn from Mr. Hunt whether his specimens were truly wild in San Miguel, or otherwise.

345. Allium subhirsutum, Linn. South Europe. Isle 1. San Miguel; Hunt coll. Not reported by any other collector besides the zealous botanist named.

The Linnean *Scilla maritima* is enumerated in Flora Azorica with the unsatisfactory habitat "ad oram maritimam," and with a duplicated number "152*" as if it were an after-thought to bring in the species at all. None of the other collectors having found this plant, it is perhaps better held doubtful for the present. Query, were the loose bulbs of the *Belladonna*, lying leafless on the sands, mistaken for the *Scilla*?

69. Smilaceæ.

346. Ruscus aculeatus, Linn. Europe. Britain. Isles 2. Miguel; Hunt coll. Terceira; Godman coll. Also in Drouet cat. 453, as seen in the same islands by his co-travellers.

347. Smilax canariensis, Willd. &c. Canaries. Isles 1, or 2. Miguel, Pico. Seubert flo. 153. 121. Watson cat. 249. Hunt coll. Drouet cat. 451 and 452? Possibly two species may be included here. My examples from Pico correspond so well with a Canary

example thus labelled by P. B. Webb, that I experience no hesitation in assigning to them the same specific name. The leaves on the examples from both these habitats are narrowly ovate-acuminate, not in the least cordate or widened at the base, strongly three-nerved with an additional faint submarginal nerve on each side. On other examples sent from San Miguel by T. C. Hunt the leaves are well described by the words of Seubert in Flora Azorica "folia late subcordato-ovata, acuminata, e basi quinque-nervia." But Seubert applies those words to specimens with the habitat of Pico; and if that is correct, both forms occur on Pico, and possibly may be states of one species. Otherwise I should have felt inclined to hold the Pico and San Miguel examples at least quasi-specifically distinct. Both Seubert and Drouet name their Pico plants "*tetragona*;" this name being given by Steudel and others as a synonym of *mauritanica*, which is certainly a different species from my own Pico specimens. I think it is different also from the San Miguel specimens; these latter perhaps being what Drouet refers to *aspera* (his no. 452) though they are quite smooth.

70. Potamaceæ.

348. Potamogeton pectinatus, Linn. Eur. Britain. Isle 1. Terceira; Seubert flo. 159. Apparently the sole record for this species in the Azores. Is the next intended?

349. Potamogeton pusillus, Linn. Europe. Britain. Isles 3. Miguel, Maria, Flores. Watson cat. 253. Hunt coll. Godman coll.

350. Potamogeton lucens, Linn. Europe. Britain. Isle 1. San Miguel; Hunt coll. Erroneously located

in "Flores (Watson)" by Drouet's Catalogue. I have seen it from San Miguel only.

351. Potamogeton polygonifolius, Pourr.

Europe. Britain.

Isles 5. Miguel, Maria, Pico, Fayal, Flores. Seubert flo. 158. Watson cat. 251 and 252. Hunt coll. Godman coll. Drouet cat. 432.

This has been variously named or misnamed. In the Flora Azorica it appeared under the name of *natans* and as inhabiting "all the islands,"—an indication of place which helps to explain what was really included under that specific name. In my own former Catalogue the same name is repeated for the plant, with the interrogative addition of "*heterophyllus* ?" for narrower-leaved examples from Flores and Corvo. Drouet's Catalogue still repeats these two names, omitting the interrogation of uncertainty after the second. I believe both names to be erroneous, having seen no example of true *natans* from the Isles; and I now refer all my own specimens, with those of Mr. Hunt and Mr. Godman, to the species known in England as *polygonifolius*. Probably this last name should also be substituted for "*Potamogeton fluitans* Roth variet. *Canariensis*" on the label no. 533 of Bourgeau's Plantæ Canarienses. At the date of my former Catalogue of Azore plants, the English botanists generally included (or, rather, confused) *polygonifolius* with *natans*; and they were doubtfully assigning narrow-leaved specimens of it from running waters to *fluitans* or *heterophyllus*. This explains my unquestioning assent to Seubert's name of *natans* for the common species of the Azores, and the doubtful assignment to *heterophyllus*, of the non flowering examples from Flores, with narrowed leaves and elongated petioles, drawn out by running water.

71. Lemnaceæ.

352. Lemna minor, Linn. Europe. Britain.

Isles . . . . ? In aquis stagnantibus; Seubert flo. 160. Tout l'archipel; Drouet cat. 437; but not marked as actually seen. Thus, it would seem that the vague indication of habitat given in the Flora, is converted in the Catalogue to the positive statement that *Lemna minor* occurs in nine islands! Such is the progress of Truth?—or of its substitute?

72. Araceæ.

353. Arum italicum, Mill. S. Europe. England.

Isle 1, or more? Miguel; Hunt coll. Fayal; by Watson cat. 250; but now supposed to have been an error. "All the islands;" Seubert flo. 161; perhaps equally an error. Godman coll.

Two segregates or subspecies pass under name of "*Arum italicum;*" and I am not prepared to assign any subordinate or specially distinctive name to the Azore specimens, which correspond with others from Madeira. Further, it is now believed that much confusion has arisen respecting the Arums of the Azores, through summer visitants finding only their fading remains. Such was my own case, in not arriving until nearly the end of May. In Flora Azorica only the *italicum* is mentioned, and it is stated to occur in all the islands. In Fayal I had observed some withered leaves which were taken to be those of *maculatum,* being smaller than the leaves of *italicum.* But on seeing the latter name alone in the Flora Azorica, I supposed a mistake on my part, and that the leaves seen might be those of the species confidently located "in all the islands." Having subsequently learned that *Arum Arisarum* is a native of the Isles, I can now better refer the

Fayal leaves to this smaller plant; thus discarding Fayal from the islands reported for *italicum*. It will remain for other observers on the spot to ascertain how much must be deducted from Seubert's averment of "all the islands." In Drouet's Catalogue we find "*Arum vulgare* Lam." given confidently as a species actually seen, growing abundantly in cultivated ground, and serving to feed the pigs. The *italicum* is also marked as a species actually seen; the *Arisarum* not marked as seen.

354. ARUM ARISARUM, Linn. South Europe.

Isles 2. Miguel; Hunt coll. Fayal; Watson cat. 250;—where the habitat is erroneously placed under *Arum italicum*, as above explained.

Colocasia antiquorum of the Flora is simply a cultivated crop, like the common Potato in Britain, and should have no place in an enumeration of the true flora of the Isles.

73. ALISMACEÆ.

355. ALISMA PLANTAGO, Linn. var. Europe. Britain.

Isle 1. Santa Maria; Godman coll. A narrow-leaved form, labelled at Kew "var. *lanceolata*."

74. JUNCACEÆ.

356. LUZULA PURPUREO-SPLENDENS, Seubert. Azores only.

Isles 5, or more. Miguel, Pico, Fayal, Flores, Corvo. All the islands; Seubert flo. 147. Watson cat. 254. Hunt coll. Godman coll. Drouet cat. 459.

It would surely be better to take up for this species the original specific name "*purpurea*" of Masson or Buch, than to adhere to a name so inconveniently long and foolish as that substituted by Seubert, in full knowledge that two earlier names *azorica* and *purpurea* had been already used for the same species. According to rule, however, Seubert's awkward compound is the name technically right; having been the first name accompanied by a figure and verbal description of the species.

357. Luzula campestris, De Cand. Europe. Britain. Isles 2. San Miguel; Hunt coll. Santa Maria (Hartung); Drouet cat. 460.

358. Juncus effusus, Linn. Europe. Britain. Isles 4. Miguel, Terceira, Fayal, Flores. Watson cat. 256. Hunt coll. Kew list of Godman coll. Drouet cat. 464.

359. Juncus glaucus, Linn. Europe. Britain. Isle 1. Santa Maria; Drouet cat. 462. Not reported by the other collectors. Said to occur in Madeira.

360. Juncus acutus, Linn. Europe. England. Isles 4, or more. Miguel, Terceira, Fayal, Flores. Most of the islands; Seubert flo. 150. 125. Watson cat. 255. Hunt coll. Godman coll. Terceira; Baron do Castello de Paiva, in Kew herbarium.

361. Juncus maritimus, Linn. Europe. Britain. Isle 1. Terceira; Seubert flo. 149. But this is not given as if it were the only island for the species, which is reported by Seubert alone.

362. Juncus capitatus, Wigel. South Europe. Isles 5. Miguel, Terceira, Fayal, Flores, Corvo. Seubert flo. 151. Watson cat. 257. Hunt coll. Godman coll. Drouet cat. 166.

363. Juncus tenuis, Willd. West Europe. America. Isles 2. Pico, Fayal. Seubert flo. 147 "*lucidus* Hochst." Watson cat. 260. Drouet cat. 463.

Two forms of this occur in Belgium, and perhaps elsewhere. In one, the branches of the panicle are elongated so as to separate the flowers, especially when advanced to fruit;—in the other, the branches are shortened and the flowers continue crowded together. This trifling difference seems to be the real distinction attempted between *tenuis* and *lucidus*, which thus divided represent states or stages of growth rather than real varieties. But the love of species-making, like the little God of Love, puts a bandage over the eyes of botanists who entertain it. The dry culms are "finely striate" in American and in European examples; and I find no difference in the "rigidity" of these latter and the Azore specimens.

364. Juncus bufonius, Linn. Europe. Britain. Isles 3. Terceira, Fayal, Flores. Seubert flo. 148. Watson cat. 259. Godman coll. Drouet cat. 469.

365. Juncus supinus, Moench. Europe. Britain. Isles 2. Miguel, Flores. Seubert flo. 152. Watson cat. 258. Hunt coll. Drouet cat. 467. "*J. uliginosus*" of the Flora, &c.

366. Juncus lamprocarpus, Ehrh. Europe. Britain. Isle 1. San Miguel; Godman coll. The specimen not much advanced towards fruit; but the name is likely correct. According to Dr. Lemann's list the same species is found in Madeira.

Another rush "*Juncus multibracteatus*, Ten.," is enumerated in Drouet's Catalogue. This is unknown to me; and the name does not appear in Bertoloni's Index.

75. Cyperaceæ.

367. Cyperus longus, Linn. S. Europe. England. Isles 5. Miguel, Maria, Terceira, Fayal, Flores. Seubert flo. 146. 153. Watson cat. 262. Hunt coll. Godman coll. Drouet cat. 472.

Perhaps some of the specimens were rightly labelled "*Cyperus badius*;" but this latter is now held by several botanists to be a form or variety of *longus*, the correctness of which I am not fully prepared to admit.

368. Cyperus esculentus, Linn. South Europe. Isles 6. Miguel, Terceira, Pico, Fayal, Flores, Corvo. Seubert flo. 145. 152. Watson cat. 263. Hunt coll. Godman coll. Drouet cat. 473 and 474, "*esculentus*" and "*aureus*."

369. Cyperus vegetus, Willd. America. Isle 1. Flores; Watson cat. 264. Two roots only were seen by myself in 1842, close by the coast town of Santa Cruz. The species was again seen by Mr. Godman in 1865, in the same island, perhaps in the same spot.

The name here used was originally taken from labels in the herbarium of Sir William Hooker and the Kew Gardens. I do not otherwise know it to be the species of Willdenow, whether that be synonymous with the American *Cyperus virens* or not so. It was likely an introduced plant in the Azores.

370. Cladium Mariscus, R. Brown. Europe. Britain. Isles 2. Miguel; Hunt coll. Flores; Watson cat. 261. Very tall in the latter island; where it was observed in one spot only, near the town or village of San Pedro, by the side of a stream and on adjacent wet ground.

371. SCIRPUS MARITIMUS, Linn. Europe. Britain. Isle 1. Terceira. Seubert flo. 142. Godman coll. Drouet cat. 479.

372. SCRIPUS SETACEUS, Linn. Europe. Britain. Isles 3. Fayal; Watson cat. 265. Flores, Corvo; Drouet cat. 480. Godman coll.

373. SCIRPUS SAVII, Seb. et Maur. Europe. Britain. Isles 5. Miguel, Terceira, Pico, Fayal, Flores. Seubert flo. 140. Watson cat. 266. Hunt coll. Drouet cat. 482.

Under name of *Isolepis Savii* in the Flora Azorica, and in Drouet's Catalogue, although the latter places its very near ally *setaceus* under the old Linnean genus *Scirpus*. To this species belongs the "*Eleocharis acicularis*" of Bourgeau's Plantæ Canarienses no. 236.

374. SCIRPUS FLUITANS, Linn. Europe. Britain. Isles 3. Miguel, Terceira, Pico. Seubert flo. 141. Watson cat. 267. Hunt coll. *Isolepis fluitans* of the Flora.

375. SCIRPUS PALUSTRIS, Linn. Europe. Britain. Isles 2. Corvo; Seubert flo. 143, under the generic name of *Eleocharis*. Flores; Watson cat. 268.

376. SCIRPUS MULTICAULIS, Sm. Europe. Britain. Isles 4. Miguel, Terceira, Fayal, Pico. Seubert flo. 144, under the generic name of *Eleocharis*. Watson cat. 269. Hunt coll. Godman coll. Drouet cat. 478.

377. CAREX SAGITTIFERA, Lowe. Madeira. Isles 5. Miguel, Terceira, Pico, Fayal, Flores. Seubert

flo. 129. 162. Watson cat. 270. Hunt coll. Drouet cat. 483. *Carex Guthnickiana*, Gay, of the Flora Azorica.

378. Carex vulpina, Linn. Europe. Britain.
Isle 1. Flores; Watson cat. 271. Apparently not observed by any other collector; and thus likely of rare occurrence.

379. Carex divulsa, Good. Europe. Britain.
Isles 3, or more. Miguel, Graciosa, Fayal. All the islands; Seubert flo. 130. Watson cat. 273. Hunt coll. Godman coll. Drouet cat. 485.

380. Carex stellulata, Good. Europe. Britain.
Isles 4. Miguel, Pico, Fayal, Flores. Seubert flo. 131 and 131 *a*. Watson cat. 272. Hunt coll.

381. Carex azorica, Gay. Azores only.
Isles 3. Miguel, Pico, Fayal. Seubert flo. 133. Watson cat. 274. Hunt coll.

382. Carex flava, Linn. Europe. Britain.
Isles 4. Miguel, Pico, Fayal, Flores. Seubert flo. 132. Watson cat. 275. Hunt coll.

The specimens from the Isles are similar to those of grassy commons in England, with small fruit and usually approximated spikes; the male spikes are mostly on short peduncles, scarcely exceeding the female spikes, occasionally on a more elongate peduncle, occasionally forming the points of the female spikes.

383. Carex lævicaulis, Hochst. Azores only.
Isles 2. Maria, Flores. Seubert flo. 134. 156. Watson cat. 276. Hunt coll.

384. CAREX RIGIDIFOLIA, Hochst. Azores only.
Isle 1. Pico; Seubert flo. 135. 160. Barão do Castello de Paiva, in Kew Herbarium?

385. CAREX HOCHSTETTERORUM, Gay. Azores only.
Isles 3. Miguel, Terceira, Fayal. Seubert flo. 136. 159. Watson cat. 278. Hunt coll. Drouet cat. 491.

386. CAREX FLORESIANA, Hochst. Azores only.
Isle 1, or more. Flores; Seubert flo. 137; only the name, without habitat expressly specified. Watson cat. 277.

387. CAREX VULCANI, Hochst. Azores only.
Isles 3, or 4. Miguel, Pico, Fayal, Flores? Seubert flo. 138. Watson cat. 277. Hunt coll. Drouet cat. 193.

388. CAREX PENDULA, Huds. Europe. Britain.
Isles 3. Miguel, Fayal, Flores. Seubert flo. 139. 155. Watson cat. 279. Hunt coll.

Dr. Seubert bestows upon this species the name of *Carex myosuroides*, Lowe. I waive a consideration of the Madeira plant; my only two examples from that island not affording sufficient ground to sustain a dogmatic opinion, that *myosuroides* of Lowe is wholly identical with *maxima* of the European botanists and *pendula* of the English botanists. I still suspect, however, that the species of the Azores, probably likewise that of Madeira, is simply a vigorous state of our British *pendula*. In Flora Azorica the alleged *myosuroides* is distinguished from the European species, by (first) its stature being three feet,—(second) its spikes being seven,—and (third) these spikes being seven or eight inches long,—(fourth) male spikes sessile, linear, not clavate,—(fifth) upper female spikes approximate,—(sixth) all male at the tips,—(seventh) utricle usually incurved. Seven characters adduced to separate a plant from something which may prove to be only itself over again! What is the value, the reality and constancy, of these alleged distinctions? At the time of writing this I have living examples of the English *pendula* before me. (First) the living stems are three and four or even five feet tall,—(second) the spikes are six or seven, counting in the one male spike,—(third) length of some of them full seven inches,—(fourth) male spikes linear or clavate in accordance with their stage of flowering,—(fifth) the upper female spikes nearer together than the lower, as usual with other many-spiked species,—(sixth) most of the female spikes are terminated by several male flowers, while some of the male spikes in turn are terminated by half-an-inch, more or less, of female flowers,—(seventh) utricle straight, that is, not yet curved in drying or by pressure. The smaller specimens of so large a plant will usually be selected for the herbarium, on the rule of convenience. But I find in my own herbarium British and European examples sufficiently vigorous to show six female spikes; some of them five or six inches long,—many female spikes terminate with male flowers,—the male spikes are clavate where the upper flowers have been in fresh perfection, and the lower flowers faded, at the time of picking the specimens, or filiform-cylindical where all are in like stage of development,—the utricles are much more curved in the dry state than they are on the living plants. In my two herbarium specimens from the Isles the female spikes are only five; and I cannot see that they clearly differ in anywise from some of the English and European examples. On the grounds here stated, I feel myself warranted in discarding the name *myosuroides* and using *pendula* as the one properly applicable to the Azore plant.

76. Gramina.

389. Anthoxanthum odoratum, L. Europe. Britain. Isles 5. Miguel, Terceira, Pico, Fayal, Flores. Seubert flo. 98. Watson cat. 286. Hunt coll. Godman coll. Drouet cat. 499.

390. Panicum sanguinale, Linn. South Europe. Isles 4. Miguel, Pico, Fayal, Flores. Seubert flo. 99. 143. Watson cat. 282. Hunt coll. Godman coll.

This is the *Digitaria sanguinalis* of my former Catalogue. It is not marked as one of the species actually seen by Drouet or his co-travellers, and yet the locality in Fayal ("environs de Horta!") has the usual sign of certainty. In the same Catalogue, Drouet enumerates a "*Panicum vaginatum* Sw." as seen by Morelet in Fayal. Does this latter name there intend the same plant?

391. Panicum Crus-galli, Linn. S.W. Europe. Isles 4. Miguel, Pico, Fayal, Flores. Watson cat. 285. Hunt coll. Godman coll. Drouet cat. 500.

392. Setaria glauca, Beauv. South Europe. Isles 5. Miguel, Terceira, Pico, Fayal, Flores. Seubert flo. 100. Watson cat. 283. Hunt coll. Godman coll. Drouet cat. 503.

393. Setaria verticillata, Beauv. South Europe. Isles 2. San Miguel; Hunt coll. Fayal; Watson cat. 284. The latter island confirmed as a habitat by Mr. Godman.

Drouet's Catalogue gives *Setaria viridis* as having been found in Terceira by M. Morelet; while the name of *verticillata* is not marked as that of a species actually seen.

The resemblance between the two is very close, and one might excusably be mistaken for the other; indeed, the mistake actually was made in the case also of Mr. Godman's specimen, which stands in the Kew list of his collections by the name of *viridis* instead of *verticillata*.

394. CYNODON DACTYLON, Pers. Europe. England.
Isles 3. Miguel, Terceira, Fayal. Seubert flo. 108. 142. Watson cat. 281. Hunt coll. Godman coll. Drouet cat. 546.

395. ELEUSINE INDICA, Gaertn. (South Europe.)
Isles 3. Miguel, Maria, Fayal. Seubert flo. 109. 137. Watson cat. 280. Hunt coll. Godman coll. Drouet cat. 547.

The variety "*brachystachya*" was found in Santa Maria by Mr. Godman, and in San Miguel by Mr. Hunt. It is sometimes mislabelled "*Dactyloctenium Egyptiacum*," a grass which has been erroneously reported from the Azores through such a misnomer.

396. ARUNDO DONAX, Linn. South Europe.
Isles 5. Miguel, Pico, Fayal, Flores, Corvo. Seubert flo. 107. Watson cat. 304. Hunt coll.

Both Seubert and Drouet describe the locality of this fine grass as being in marshes. In the islands visited by myself I do not now recollect ever to have seen it in such situations. It was used there to form fences, which are likely a serviceable protection to the crops against the sweeping winds from the surrounding ocean. The lines of it thus made on the hill-sides became one of the peculiar features of the landscape, as looked up to from the surface of the ocean. Perhaps introduced to the Isles for that useful purpose.

397. Agrostis alba, Linn. Europe. Britain. Isles 4. Miguel, Fayal, Flores, Corvo. Watson cat. 294. Hunt coll. Godman coll. Drouet cat. 506.

398. Agrostis verticillata, Vill. Europe. Britain. Isles 4. Miguel, Terceira, Fayal, Flores. Seubert flo. 101. 149. Watson cat. 295. Hunt coll. Drouct cat. 507.

Generally deemed to be a variety of the preceding species, which I am not prepared to deny, although not wholly satisfied on the point.

399. Agrostis vulgaris? Europe. Britain. Isle 1, or 2. Summit of the Peak of Pico, 7600 feet; Watson cat. 296. Perhaps the same grass sent also from San Miguel by Mr. Hunt. The name remains still highly uncertain.

400. "Deyeuxia cæspitosa, Hochst. msc." Azores only. Isles 4. Miguel, Pico, Fayal, Flores. Seubert flo. 105. 188. Watson cat. 298. Hunt coll. Drouet cat. 513.

The "*Agrostis pallida*?" of my former Catalogue, no. 298, very likely belongs to the present species. There is certainly something within the glumes which Dr. Seubert may be warranted in designating as "the plumose rudiment of another flower."

401. "Deyeuxia azorica, Hochst. msc." Azores only. Isles 2, or 4. Terceira, Fayal; Seubert flo. 106. Terceira, Pico, Fayal, Corvo; Drouet cat. 514.

This remains imperfectly known to me, if truly known at all. It may be my no. 293, which Dr. Lemann afterwards confidently referred to *Piptatherum multiflorum*; but concerning which I yet reach no definite opinion or knowledge. In fact, I have given but little attention to the Grasses, and scarcely like to express any positive opinion in cases of confused names or doubtful diagnosis. (See also *Deschampsia argentea*, below.)

402. Gastridium australe, Beauv. Europe. England. Isles 5. Miguel, Maria, Pico, Fayal, Flores. Seubert flo. 102. 144. Watson cat. 289. Hunt coll. Godman coll. Drouet cat. 509 and 510.

I know not how he would distinguish the plants themselves, but M. Drouet marks the names *australe* and *lendigerum* as those of two different species, both actually seen in the Isles. They are simply synonymous.

403. Polypogon monspeliensis, Desf. S. Eur. England. Isles 5. Miguel, Terceira, Pico, Fayal, Flores. Seubert flo. 104. 189. Watson cat. 288. Hunt coll. Godman coll. Drouet cat. 512.

404. " Polypogon maritimus, Willd." South Europe. Isles 2, or more. Several of the islands; Seubert flo. 103. 133. Terceira and Fayal; Godman collection, according to the Kew list. San Miguel and other islands; Drouet cat. 511.

This grass is unknown to me; perhaps I confuse it with the *P. monspeliensis*, a name which was used on my own labels, and also perhaps on all those written for specimens from Mr. Hunt. The specimens collected by Mr. Godman were labelled at Kew as *maritimus*; by me they would have been labelled as *monspeliensis*. Authors generally appear to see two different species to represent the two names.

By the second no. to each in Seubert's Flora, it would seem that both were distributed in Hochstetter's collection.

405. Lagurus ovatus, Linn. South Europe.

Isles 3, or more. Miguel, Fayal, Flores. All the islands; Seubert flo. 111. 138. Watson cat. 287. Hunt coll. Godman coll. Drouet cat. 531. Flores; Barão do Castello de Paiva in Kew Herbarium.

406. "Piptatherum multiflorum, Beauv." S. Europe.

Isle 1, or more. Corvo; Watson cat. 293 (*fide* C. Lemann) under *Deyeuxia azorica.* Perhaps sent also from San Miguel by Mr. Hunt. (See London Journal of Botany, vol. vi. p. 393.)

For the grounds of uncertainty about this plant or name, see the *Deyeuxia azorica* above.

407. Aira caryophyllea, Linn. Europe. Britain.

Isles 5. Miguel, Terceira, Pico, Fayal, Flores. Seubert flo. 110. 148. Watson cat. 297. Hunt coll. Godman coll. Drouet cat. 518. Terceira is named on authority of Mr. Godman's specimens; adding a fifth island to those before on record.

408. Deschampsia argentea, Lowe. Madeira.

Isles 2. Miguel, Flores. Watson cat. 299. Hunt coll. Drouet cat. 520, not marked as actually seen.

Possibly this may be one of the two *Deyeuxiæ* of the Flora Azorica; being otherwise omitted from that work, although a conspicuous grass not likely to escape the notice of a collector. M. Drouet fails to mark it as a grass actually seen in the Isles, and yet he describes it like an eye-witness as being "very abundant on the mountain pastures, of which this grass forms one of the principal elements."

409. AVENA HIRSUTA, Roth. South Europe.

Isles 4. Miguel, Maria, Terceira, Fayal. Seubert flo. 112. 150. Watson cat. 303. Hunt coll. Drouet cat. 523. "*Avena barbata*, Brot. lusit. 108."

Perhaps *Avena fatua* also occurs in San Miguel; but I am not certain as to the species of my solitary young example.

410. AVENA BREVIS, Roth. Mid Europe.

Isles 2. Terceira; Seubert flo. 113. Fayal; Drouet cat. This is not known to me, and will thus rest on the authorities cited.

411. ARRHENATHERUM AVENACEUM, Beauv.
Europe. Britain.

Isles 3. Miguel, Fayal, Flores. Watson cat. 302. Hunt coll. Godman coll. Drouet cat. 522. "*Avena elatior*" of the Catalogues cited.

412. HOLCUS LANATUS, Linn. Europe. Britain.

Isles 6. Miguel, Terceira, Graciosa, Pico, Fayal, Flores. Watson cat. 300. Hunt coll. Godman coll. Drouet cat. 497. Pico; Barão do Castello de Paiva in Kew Herbarium.

413. HOLCUS RIGIDUS, Hochst. Azores only.

Isles 4. Miguel, Pico, Fayal, Flores. Seubert flo. 96. 146. Watson cat. 301. Hunt coll. Godman coll. Drouet cat. 496.

Very likely *Holcus mollis* may not have occurred in the Isles. That name was inadvertently used in my Notes for the present quite distinct species. Drouet cites Hartung for a Fayal habitat of *H. mollis*, which may be due to a like error.

414. Koeleria phleoides, Pers. South Europe.
Isles 2. Miguel, Fayal. Seubert flo. 119. 132. Watson cat. 292. Hunt coll. Drouet cat. 535.

415. Poa annua, Linn. Europe. Britain.
Isles 4. Miguel, Maria, Terceira, Fayal. Watson cat. 315. Hunt coll. Godman coll. Drouet cat. 527.

416. Poa trivialis, Linn. Europe. Britain.
Isles 5. Miguel, Maria, Pico, Fayal, Flores. Watson cat. 314. Hunt coll. Godman coll. Drouet cat. 529.

Drouet's Catalogue also enumerates *Poa pratensis*, no. 528, as found in Fayal by Morelet.

417. Poa Eragrostis, Linn. South Europe.
Isles 3. Miguel; Hunt coll. Terceira (Morelet); Drouet cat. 532. Pico; Watson cat. 316.

The old Linnean name is retained for this species partly because I really do not know (and do not care) under which of the segregated species of *Eragrostis* it would be placed by the semi-species discriminators. Not mentioned in the Flora Azorica. In my former Catalogue it is enumerated under the name here retained for it. In Drouet's Catalogue it is made into two separate species, namely, *Eragrostis megastachya* and *poæoides*, both alleged to have been found in Terceira, by M. Morelet.

418. Poa rigida, Linn. Europe. Britain.
Isles 3, or more. Miguel, Terceira, Fayal. Seubert

flo. 116. 135. Watson cat. 317. Kew list of Godman coll. Drouet cat. 530.

This and the next are kept under the Linnean genus *Poa* for sake of uniformity; that arrangement having been adopted in the Flora and in the two Catalogues cited. In English Floras it goes also under the generic names of *Glyceria* and *Sclerochloa*; and it is the *Scleropoa rigida* of Griesbach.

419. Poa loliacea, Huds. Europe. Britain.

Isles 3. Miguel, Terceira, Fayal. Seubert flo. 115. 136. Hunt coll. Drouet cat. 526, "tout l'archipel";—a record likely made at random.

Considerable confusion has arisen through the difficulty of uniting this grass into a genus with other species. It has been put under various genera, some of them sufficiently dissimilar among themselves; namely, *Triticum, Brachypodium, Catopodium, Festuca, Sclerochloa, Scleropoa, Glyceria*; the specific name also undergoing varations accordingly.

420. Briza maxima, Linn. South Europe.

Isles 7. Miguel, Maria, Terceira, Pico, Fayal, Flores, Corvo. Seubert flo. 117. 147. Watson cat. 305. Hunt coll. Godman coll. Drouet cat. 533.

421. Briza minor, Linn. Europe. England.

Isles 4. Miguel, Terceira, Fayal, Flores. Seubert flo. 118. 146. Watson cat. 306. Hunt coll. Godman coll. Drouet cat. 534.

422. Triodia decumbens, Beauv. Europe. Britain.

Isle 1. San Miguel; Hunt coll. It is also enumerated in Drouet's Catalogue, but not marked there as a species actually seen in the Isles; so that M. Drouet likely adopted it from a manuscript list of Mr. Hunt's further additions (unpublished) to the flora of the Isles.

423. Cynosurus echinatus, Linn. South Europe. Isles 4. Miguel, Terceira, Pico, Fayal. Seubert flo. 119*. 139. Watson cat. 291. Hunt coll. Godman coll. Drouet cat. 536.

424. Cynosurus cristatus, Linn. Europe. Britain. Isles 2. Santa Maria (Hartung), Pico (Morelet); Drouet cat. 537. Not found by other collectors.

This grass is not found in Madeira or the Canaries, according to my compiled floral lists for those islands; nor is it in the Flora of Algeria. It would on that account be desirable to have a confirmation of the reported habitats in the Azores, by somebody who makes botany his more special study.

425. Festuca bromoides, Linn. Europe. Britain. Isles 4, or more. Miguel, Terceira, Pico, Fayal. Almost all the islands; Seubert flo. 122. 140. Watson cat. 311. Hunt coll. Kew list of Godman coll. Drouet cat. 541.

I follow the Flora Azorica in adopting this name, though not with absolute trust in its correctness. My specimens are young, and in them the stem is leafy up to the rather long panicle; thus giving to the specimens much the appearance of the grass known in England by the name of *Pseudo-Myurus*.

426. Festuca jubata, Lowe. Madeira. Isles 2, or 3. Fayal, and probably Corvo; Watson cat.

313. Also in Pico, if it be the same with *Festuca glauca* var. *longearistata* of Seubert's Flora, no. 120.

In Flora Azorica, the *Festuca glauca* is located on coast-rocks in Fayal and Pico. My specimens were picked inland on the mountains, unless memory now deceives.

427. FESTUCA PETRÆA, Guthnick. Azores only.

Isles 3, or more. Miguel, Terceira, Fayal. All the islands; Seubert flo. 121. 131. Watson cat. 312. Hunt coll. Drouet cat. 540. On coast cliffs.

428. FESTUCA ELATIOR, Linn. Europe. Britain.

Isles 2. San Miguel; Hunt coll. Santa Maria (Hartung); Drouet cat. 542.

Mr. Hunt's specimens belong to the plant known in England as *elatior* or *arundinacea*, by their short and almost ovate spikelets,—not to the *pratensis* distinguished by its linear spikelets. Much confusion and cross-naming have occurred between our English species.

429. BROMUS MADRITENSIS, Linn. S. Eur. England.

Isles 3. Miguel, Fayal, Flores. Seubert flo. 124. 141*b*. Watson cat. 307. Hunt coll. Godman coll. Drouet cat. 546.

430. BROMUS RUBENS, Linn. South Europe.

Isles 2. San Miguel; Seubert flo. 123. 141. Terceira; Kew list of Godman coll.

There are specimens from Mr. Hunt, which likely belong to this species, but I abstain from directly citing them, not being familiar with the species itself. From Mr. Hunt I have also received one specimen which may be *Bromus maximus*.

431. Bromus mollis, Linn. var. Europe. Britain. Isles 3. Miguel, Fayal, Pico. Watson cat. 308. Hunt coll. Spikelets less ovate, and more pubescent than seen in the English examples of the species.

432. Hordeum murinum, Linn. Europe. Britain. Isles 3. Miguel, Terceira, Fayal. Seubert flo. 128. Watson cat. 319. Hunt coll. Kew list of Godman coll. Drouet cat. 555.

433. Triticum repens, Linn. Europe. Britain. Isles . . . ? "Hab. ad vias et in graminosis;" Seubert flo. 126. Introduced? Seemingly not found by any other collector.

This common grass of Europe is reported also for Madeira and Canaries; thus indirectly suggesting a likelihood of its occurrence in the Azore Isles. But could the next species, not recorded in the Flora by Seubert, have been mistaken for *Triticum repens*?

434. Brachypodium sylvaticum, Beauv. Eur. Britain. Isles 4. Miguel, Terceira, Pico, Fayal. Watson cat. 310. Hunt coll. Godman coll. Drouet cat. 543.

435. Brachypodium distachyum, Beauv. S. Europe. Isles 4. Miguel, Maria, Terceira, Pico. Seubert flo. 127. 134, under name of *Triticum ciliatum*. Watson cat. 309. Hunt coll. Drouet cat. 534.

436. Lolium perenne, Linn. Europe. Britain. Isles 4. Miguel; Hunt coll. Terceira and Pico; Drouet cat. 549. Fayal; Godman coll. Flores; Dr. Mackay! Introduced?

437. Lolium multiflorum, Lamarck? South Europe. Isles 5. Miguel, Terceira, Pico, Fayal, Flores. Seubert flo. 125. Watson cat. 318. Hunt coll. Godman coll. Drouet cat. 551 and 552.

Drouet's Catalogue enumerates *Lolium italicum* as a species actually seen in Terceira and Pico, and *Lolium multiflorum* as another species not seen by himself or friends. Do not the two names used really intend one single species, as in so many other similar cases? *Lolium arvense* is also enumerated, as having been seen in Santa Maria by Hartung;—perhaps imported with grain from Portugal, if that grass were really seen.

438. Gaudinia geminiflora, Gay. Azores only. Isles 2. Miguel, Fayal. Seubert flo. 114. 151. Watson cat. 290. Hunt coll. Drouet cat. 525.

439. Nardus stricta, Linn. Europe. Britain. Isle 1. San Miguel; Hunt coll. No other collector appears to have found this grass; one not likely to have been mistaken for anything else on record for the Isles.

77. Filices.

440. Dicksonia Culcita, Herit. Madeira. Spain? Isles 5. Miguel, Terceira, Pico, Fayal, Flores. Seubert flo. 88. 172. Watson cat. 340. Hunt coll. Godman coll. Drouet cat. 593.

I learn from my nephew Major Wakefield that this fine fern has been found (1869) in the South of Spain; is it native there?

441. Hymenophyllum tunbridgense, Sm. Europe. Britain. Isles 5. Miguel, Terceira, Pico, Fayal, Flores. Seu-

bert flo. 89. 187. Watson cat. 342. Hunt coll. Godman coll. Drouet cat. 594.

442. HYMENOPHYLLUM UNILATERALE, Bory.
Europe. Britain.
Isles 4. Miguel, Terceira, Flores, Corvo. Watson cat. 343. Hunt coll. Drouet cat. 595. *H. Wilsoni* in the Catalogues cited.

443. TRICHOMANES SPECIOSUM, Willd. Spain? Britain.
Isles 5. Miguel, Terceira, Pico, Fayal, Flores. Seubert flo. 90. 181 and 185. Watson cat. 341. Hunt coll. Godman coll. Drouet cat. 596.

Drouet's Catalogue also enumerates "*Trichomanes canariense*" (Davallia canariensis) on authority of a Catalogue of Plants in a Lisbon botanic garden, edited by two Portuguese botanists, Gomes and Beirao. This may be an error of habitat in the Catalogue mentioned; at any rate, a Fern so peculiar in appearance, could hardly be overlooked by all the English, French, and German collectors who have visited the Isles, unless extremely local.

444. CYSTOPTERIS FRAGILIS, Bernh. Europe. Britain.
Isles 5. Miguel, Terceira, Pico, Fayal, Flores. Seubert flo. 87. 179. Watson cat. 321. Hunt coll. Godman coll. Drouet cat. 592.

445. ADIANTUM CAPILLUS-VENERIS, Linn.
S. Europe. England.
Isle 1. San Miguel; Seubert flo. 72. 168 and 169. Hunt coll. Godman coll. Drouet cat. Frequent in San Miguel, according to a letter from Mr. Hunt, although by some oversight he had omitted to send specimens of it.

446. PTERIS ARGUTA, Aiton. Portugal. "Corfu."
Isles 4. Miguel, Pico, Fayal, Flores. Seubert flo. 74.

184. Watson cat. 337. Hunt coll. Kew list of Godman coll. Drouet cat. 571.

447. Pteris aquilina, Linn. Europe. Britain.
Isles 5. Miguel, Maria, Pico, Fayal, Flores. Seubert flo. 73. Watson cat. 338. Hunt coll. Godman coll. Drouet cat. 570.

448. Lomaria Spicant, Desv. Europe. Britain.
Isles 5, or more. Miguel, Terceira, Pico, Fayal, Flores. Seubert flo. 75. 177. Watson cat. 336. Hunt coll. Godman coll. Drouet cat. 572 and 573.

Drouet's Catalogue makes enough of this Fern, enumerating "*Blechnum boreale*" and "*Blechnum Spicant*" as two different species, both actually seen in the Isles! Apparently, he was not aware that the two names are simply synonyms for a single species. The difficulty is now to understand how he could see the two imaginary species, and not see that they were one only.

449. Woodwardia radicans, Swartz. South Europe.
Isles 4. Miguel, Pico, Fayal, Flores. Seubert flo. 83. Watson cat. 339. Hunt coll. Kew list of Godman coll. Drouet cat. 583.

450. Asplenium palmatum, Swartz. Portugal.
Isles 5. Miguel, Terceira, Pico, Fayal, Flores. Seubert flo. 76. 178. Watson cat. 354. Hunt coll. Godman coll. Drouet cat. 574.

In the 'Synopsis Filicum' of Hooker and Baker, this is held to be the same with *Asplenium Hemionitis*.

451. Asplenium anceps, Solander. Europe. Britain.
Isles 5. Miguel, Maria, Pico, Fayal, Flores. Seubert flo. 78. 174. Watson cat. 326. Hunt coll. Godman coll. Drouet cat. 577.

This is scarcely more than another name for the *Asplenium Trichomanes* of Britain. Primarily it intends a luxuriant variety, which is perhaps the prevalent form in the more southern latitudes, and quite absent towards the northern limits of the species. In Surrey, I have seen fronds of *Trichomanes* a foot in length and the pinnæ more than half-an-inch;—larger fronds than any in my herbarium from the Azores, Madeira, or Canaries. These islands have not only special "forms," but several forms; just as occurs in Britain.

452. Asplenium monanthemum, Smith.
Madeira. Canaries.
Isles 4. Miguel, Pico, Fayal, Flores. Seubert flo. 80. 175. Watson cat. 325. Hunt coll.

453. Asplenium marinum, Linn. Europe. Britain.
Isles 3, or more. Miguel, Fayal, Flores. "All the islands;" Seubert flo. 79. 173. Watson cat. 327. Hunt coll. Godman coll. Drouet cat. 578.

454. Asplenium lanceolatum, Huds.
S. Europe. England.
Isles 3. Miguel, Fayal, Flores. Watson cat. 329. Hunt coll. Godman coll.
Somewhat remarkable for this species to have been omitted from the Flora Azorica; being sufficiently frequent and obvious in the Isles. Perhaps it was passed by as one of the varieties of the proteiform *Adiantum-nigrum*. Apparently, it has been seen or noticed by the English collectors only.

455. Asplenium Adiantum-nigrum, Linn.
Europe. Britain.
Isles 4. Miguel, Pico, Fayal, Flores. Seubert flo. 77.

176. Watson cat. 328. Hunt coll. Kew list of Godman coll. Drouet cat. 575. (Resembles the *A. productum* of Lowe.)

456. ATHYRIUM FILIX-FŒMINA, Swartz. Eur. Britain. Isles 5. Miguel, Maria, Terceira, Pico, Flores. Watson cat. 330. Hunt coll. Kew list of Godman coll. Drouet cat. 580, under the generic name of *Asplenium*.

457. ATHYRIUM UMBROSUM, Presl. Madeira. Canaries. Isles 3. Miguel, Pico, Flores. Seubert flo. 81. 183. Watson cat. 332. Hunt coll. Kew list of Godman coll. Drouet cat. 581.

Under the name of *Allantodea umbrosa* in the Flora and two Catalogues cited. Another species, "*Allantodea axillaris*" appears in Seubert's Flora and Drouet's Catalogue. This is not known to me; but by Milde it is held to be a variety of *Filix fœmina*; which may help to explain the omission of this last fern from the Flora Azorica.

458. SCOLOPENDRIUM VULGARE, Smith. Eur. Britain. Isles 3, or more. Miguel, Fayal, Flores. Almost all the islands; Seubert flo. 84. Watson cat. 323. Hunt coll. Kew list of Godman coll. Drouet cat. 584.

459. ASPIDIUM ANGULARE, Willd. Europe. Britain. Isles 3, 5, or more. Miguel, Fayal, Flores. All the islands; Seubert flo. 86. 180. Watson cat. 334. Hunt coll. Drouet cat. 587. Also reported for Terceira and Pico, if "*Aspidium aculeatum*" of Drouet's Catalogue, no. 588, is only another name for the present species.

To my apprehension this is a species quite as distinct from *Aspidium lobatum*, as the *dilatatum* and *æmulum*, or the *Filix-mas* and *rigidum* are distinct from each other. Yet it is constantly confused with *lobatum*, apparently in part so confused through misapplications of the name *aculeatum,* used indiscriminately for either or for both. On the Continent, the *aculeatum* seems to be represented more frequently by specimens of *angulare.* In Britain, that mistake is not so often made; for we usually take the most divided states of *lobatum* to represent the *aculeatum.* I have not seen true *lobatum* or *its* variety *aculeatum* from the Isles.

460. ASPIDIUM FILIX-MAS, Swartz. Europe. Britain. Isles 3. Miguel; Hunt coll. Fayal and Flores; Drouet cat. 589, under the generic name *Polystichum.*

I avoid use of the sub-generic names *Lastrea* and *Polystichum* because they are applied to the species differently on the Continent and in Britain; that group which we name *Lastrea,* is named *Polystichum* by some of the Continental authors.

461. ASPIDIUM DILATATUM, Willd. Europe. Britain. Isles 1, or 4. Miguel; Hunt coll. Maria; Drouet cat. 568. Fayal and Flores; Drouet cat. 590.

Although the "*Lastrea multiflora*" of Newman and the "*Polystichum dilatatum*" of De Candolle are placed far apart in the list of Ferns in Drouet's Catalogue, as shown by their respective nos. above cited, I presume them to be simply two synonyms for *Aspidium dilatatum.* Besides the two species and genera made out of those two names, Drouet's Catalogue has likewise a *Polystichum tanacetifolium* (De Candolle) given as a third species. On the contrary, *Nephrodium Fœnisecii* (Lowe) is entered in the Catalogue as the name of a species not seen in the Isles by M. Drouet or his friends; so that possibly the "*tanacetifolium,*" marked as having been actually seen, may really intend Lowe's fern, treated below under its earlier specific name *æmulum*. M. Drouet makes three (or four) species, placed under three different genera, out of the *dilatatum* and *æmulum* as here treated.

462. Aspidium æmulum, Swartz. Britain. Madeira. Isles 4. Miguel, Pico, Fayal, Flores. Seubert flo. 85. 181. Watson cat. 331. Hunt coll. Godman coll.

Under name of '*Nephrodium Fœnisecii* Lowe' in the Flora and Catalogue cited. Milde states that examples of *spinulosum* and *dilatatum* were commingled with this under Hochstetter's no. 181. I have not seen true *spinulosum* (as understood in England) from the Isles. Milde himself has failed to extricate the two species, although the living plants ought never to be confused. Their manner of growth is quite dissimilar; and the leafless crowns of the two are readily distinguished from each other in winter, equally as the living leaves in summer.

463. Aspidium molle, Swartz. Madeira. Canaries. Isles 3. Miguel, Fayal, Flores. Watson cat. 333. Hunt coll. Kew list of Godman coll. Drouet cat. 586.

464. Polypodium vulgare, Linn. Europe. Britain. Isles 3, or more. Miguel, Fayal, Flores. All the islands; Seubert flo. 71. 171. Watson cat. 320. Hunt coll. Godman coll. Drouet cat. 567.

465. Gymnogramme Lowei, Hook. et Grev. Madeira. Africa. Isles 2. Miguel; Hunt coll. Fayal; Drouet cat. 566. Kew list of Godman coll.

Said to be the same with *Gymnogramme Totta*, of Schlectendal, a widely distributed Fern in latitudes southward from the Azore Isles. See 'Synopsis Filicum,' by Hooker and Baker.

466. GYMNOGRAMME LEPTOPHYLLA, Swartz. S. Europe. Isles 4. Miguel, Maria, Terceira, Fayal. Seubert flo. 70. 166. Watson cat. 322. Hunt coll. Godman coll. Drouet cat. 565.

467. ACROSTICHUM SQUAMOSUM, Swartz. Madeira. Africa. Isles 5. Miguel, Terceira, Pico, Fayal, Flores. Seubert flo. 69. 170. Watson cat. 335. Hunt coll. Godman coll. Drouet cat. 564.

468. OSMUNDA REGALIS, Linn. Europe. Britain. Isles 4. Miguel, Terceira, Fayal, Flores. Seubert flo. 91. 167. Watson cat. 344. Hunt coll. Kew list of Godman coll. Drouet cat. 598.

469. OPHIOGLOSSUM VULGATUM, Linn.? Europe. Britain. Isle 1. Flores; Watson cat. 345. Godman coll., from the same island. Terceira? (See the next.)

470. OPHIOGLOSSUM LUSITANICUM, Linn. South Europe. Isles 2. Miguel; Hunt coll. Terceira; Seubert flo. 92. 165, under name of *Ophioglossum polyphyllum* Braun;—unless this latter is synonymous with some variety of *vulgatum*.

It is truly difficult to decide whether a single species only or two good species exist in the Isles. The Flora Azorica assigns the specimens from Terceira, seen by Dr. Seubert, to *O. polyphyllum*. Braun's species is not known to me otherwise; and I have not seen any specimens from Terceira. But I have not the least hesitation in assigning Mr. Hunt's numerous specimens from San Miguel to the *lusitanicum*, as represented in my herbarium by specimens from Guernsey, France, Portugal, Spain, Italy, and Algeria. The examples from Flores, where an

Ophioglossum was found by Mr. Godman and myself, are much larger; and to me they appear inseparable from *O. vulgatum* of Britain, Europe, and North America. If this were all, our way would seem clear to furnish the Isles with two different species, *vulgatum* and *lusitanicum*; the *polyphyllum* united with one of them. Unfortunately for the clearness, however, my Madeira specimens from Dr. Charles Lemann (by him labelled "*lusitanicum*") stand exactly between the *lusitanicum* of San Miguel and the *vulgatum* of Britain; while the Flores specimens might also be said to continue the intermediate series, being nearer to *vulgatum* in size and form, and yet slightly shading off towards *lusitanicum*. I kept several plants of the diminutive *lusitanicum* of Guernsey some few years in a flower-pot. They flourished in a cold frame in winter, retaining their small size and narrow fronds; fading away in spring before *vulgatum* had attained its full expansion in the open ground. The plants were all lost by an accident after the second or third winter; so that observation of them was not continued long enough to make sure they would never become more like the *vulgatum* of Flores, or that of Britain.

78. Lycopodiaceæ.

471. Lycopodium cernuum, Linn. South America, etc. Isles 2. Miguel, Terceira. Seubert flo. 94. 163. Hunt coll. Godman coll. Drouet cat. 560.

472. Lycopodium complanatum, Linn. Mid Europe. Isle 1. San Miguel; Hunt coll. Drouet cat. Kew Herbarium from the Barão do Castello de Paiva.

473. Lycopodium suberectum, Lowe. Madeira. Isles 5. Miguel, Terceira, Pico, Fayal, Flores. Seu-

bert flo. 93. 164. Watson cat. 347. Hunt coll. Godman coll. Drouet cat. 561.

This is often seen hanging down like a chain from rocks and steep banks; and in such situations it would remind the English observer more of *clavatum* than of *Selago,* as these are seen on the mountain moors of Britain. And yet here and there on grassy slopes small erect examples could be found, which came so near to our British *Selago* in general appearance, as scarcely to be separated from the latter, except by the spinulose teeth along the margins of the leaves. And one of my examples, picked high on Pico, by wanting these processes, becomes absolutely undistinguishable from *Selago,* for which it is given below, supported also by Drouet's additional habitat of Terceira. In Flora Azorica, Lowe's species is included under the specific name of *Selago* "var. *spinulosum* Spring." It is so placed also in Milde's 'Filices Europæ et Atlantidis,'—a work in which there is the same tendency to an excessive aggregation of species, which so much lowers the scientific value and serviceableness of Sir William Hooker's works on the same group of plants.

474. Lycopodium Selago, Linn. Europe. Britain. Isles 2. Pico; Watson cat. 346. Terceira; Drouet cat. 559. See this mentioned under the preceding species.

475. Selaginella Kraussiana, Kunze.
Madeira. S. Africa.
Isles 5, or more. Miguel, Terceira, Pico, Fayal, Flores. All the islands; Seubert flo. 95. Watson cat. 348. Hunt coll. Godman coll. Drouet cat. 563.

This species has always passed in our gardens and greenhouses for *Lycopodium* (*Selaginella*) *denticulatum.* Hence, I too readily followed the example of Seubert's Flora in thus labelling my own Azore specimens. Those reserved for my herbarium belong to *Kraussiana*; and it may be deemed highly probable that such was the fact with all the other specimens distributed by me, whether collected by myself or by Mr. Hunt. I assume this to be equally true also of the *Selaginella* recorded as *denticulata* in the works by Seubert and Drouet. If so, the true *S. denticulata* should for the present be held as not ascertained to occur in the Isles; where, however, it may be thought likely enough to occur, on the inference that a species of Madeira and Spain is thus likely to be Azoric also.

79. Isoetaceæ.

476. Isoetes azorica, Durieu (ex Milde). Azores only? Isle 1. Corvo; Watson cat. 349, under name of *Isoetes lacustris*.

Growing in shallow water in the Isle of Corvo; apparently the only ascertained habitat for the species. I know not how Dr. Milde got his printed habitat of "Flores (Watson)." On present recollection, I possessed extremely few examples of this plant; and the only one reserved for my own herbarium was afterwards sent to M. Durieu, at the request of M. Gay of Paris.

80. Equisetaceæ.

477. Equisetum maximum, Lam. Europe. Britain. Isles 3. Miguel; Hunt coll. Fayal, Flores; Watson cat. 350. Also marked in Drouet's Catalogue as a plant actually seen in Flores.

This is the *Equisetum Telmateia* (Ehrh.) of Drouet's Catalogue, no. 556, and the *E. fluviatile* (Smith) of my own former Catalogue, no. 350. These are not names falsely applied to the species of the Isles, but simply synonyms for the same plant.

478. Equisetum incanum, Vauch. Canaries, etc.
Isle 1. San Miguel; Hunt coll. Godman coll.

Mentioned in my Supplementary List under the erroneous name of *limosum*. An after supply of better specimens enabled me to correct that misnomer; and the later-dated specimens from Mr. Hunt were thus rightly labelled as *incanum*. The plant intended under this latter name is now held a variety of *Equisetum elongatum*, as I first learned from my much esteemed friend in botany, Mr. J. G. Baker. Dr. Milde places it under a very aggregate species "*ramosissimum* R. Desf.," distributed over the earth, from Faroe and Siberia northwards, to the Cape of Good Hope and corresponding latitudes in South America. In this, as in numerous other similar cases, the geographical distribution is made to depend artificially on the arbitrary combinations or severances of forms, more or less closely allied to each other.

III. GENERAL REMARKS.

The Catalogue gives only a scanty flora for a group of nine islands, situate so far southward as latitude 37–40, and one of them near forty miles across by its longest diameter. But small as the ascertained flora may appear, it seems now unlikely that there can be any considerable further addition to the list of flowering plants and ferns; the collections of the last four travellers there (Drouet, Morelet, Hartung, Godman) having together added only about a score of species to those previously reported. Indeed, if it were sought to make an exclusive list of the true natural flora, it seems reasonable to suppose that the number of species would decrease rather than increase; any further native plants yet to be added to such a list, being more than equalled by those which ought to be excluded from present lists as being almost certainly introductions into the Isles through human agencies. The 478 species which are enumerated in the foregoing Catalogue, must be held to include among them many plants which have been carried to the Isles from Europe and elsewhere; while too likely some of them are erroneous determinations; and notwithstanding that several others have been already left out of the list, as very likely belonging to one or other of the two categories.

Five collections have been combined to form the foregoing Catalogue; those of Hochstetter (Flora Azorica), Watson, Hunt, Godman, Drouet with his co-travellers of 1858. As the reported species are adopted into the Catalogue, those of the several collections count up to the following numbers:—

Hochstetter, in Seubert's Flora . . .	289
Watson, 338; adding thereto . . .	103
Hunt, resident, 375; adding	67
Drouet, Morelet, Hartung, 322; adding	13
Godman, 256 or more; adding . . .	6
	478 species.

Some of the plants reported by M. Drouet or his fellow travellers were not accepted into the regular list, and some few of those which do appear there have been admitted with much distrust; for instance, *Moehringia muscosa* and *Potentilla verna*. But taking M. Drouet's additions at a full dozen, with Mr. Godman's undoubted half-dozen, we get scarcely a score of additions to the species already on record for the Isles before 1850. These are under five per cent. of novelties; and if they give us a fair measure of what may be expected from future visitors, the presently ascertained flora of the Isles may be held approximately full, so far as the flowering plants and ferns are concerned.

Against this inference, however, we must set the facts, that one only of the collectors named (Mr. Hunt) was a resident, and that the two islands of San Jorge and Graciosa have been scarcely examined, and also Pico perhaps very little examined, except along the line of ascent from the usual landing-place on the coast to the summit of its peak.

With the exceptions presently to be mentioned, it is believed that examples of all the plants enumerated in the Catalogue have passed under my own eye. And with extremely few additional exceptions, it may be said further that all of them are represented by specimens fastened to papers in my own herbarium. The following 38 species are retained in the Catalogue on faith of other records or testimony only; actual examples not having been seen by myself:—

1. Ranunculus Flammula. Reported by M. Morelet.
7. Nigella arvensis. Introduced through gardens.
15. Fumaria officinalis. A plant likely to occur.
16. Fumaria micrantha. Alien or error of name?
19. Nasturtium flexuosum. An alleged new species.
24. Sisymbrium Irio.
28. Senebiera Coronopus.
50. Moehringia muscosa. Probably a misnomer.
54. Spergularia marina. If distinct from nos. 53 and 55.
66. Malva rotundifolia. Apparently a misnomer.
71. Geranium rotundifolium. Probably correct.
83. Medicago lupulina.
93. Trifolium lappaceum.
103. Lotus corniculatus.
107. Lotus creticus.
128. Potentilla anserina. An erroneous record?
129. Potentilla reptans. Some form of Tormentilla?
131. Potentilla verna. A misnomer?

134. Poterium Sanguisorba.
153. Ammi majus.
158. Pimpinella dichotoma.
162. Angelica montana.
170. Galium Mollugo. Probably some error.
242. Datura Stramonium.
245. Verbascum spurium. An uncertain species.
260. Digitalis purpurea. Reported by two observers.
281. Myosotis stricta. Suspected to be M. collina.
290. Anagallis tenella. Confirmation very desirable.
303. Chenopodium rubrum. Verification desirable.
348. Potamogeton pectinatus. Was this P. pusillus?
352. Lemna minor.
359. Juncus glaucus.
360. Juncus maritimus.
384. Carex rigidifolia.
401. Deyeuxia azorica. Is this Deschampsia argentea?
410. Avena brevis. The species not known to me.
424. Cynosurus cristatus. An introduced grass?
433. Triticum repens.

Unless it be the *Moehringia muscosa* and *Potentilla verna*, none of these 38 plants are so very unlikely to occur in the Isles as to warrant a total rejection of any of them until further investigation shall have been made; although I do much suspect that about half of them are errors of name or of record somehow. For the present, they must be taken into our arithmetical or statistical reckonings, as recorded constituents of the flora.

The nine islands arrange naturally into three groups. The large island of San Miguel, with the islet of Santa Maria, will constitute the most southerly and easterly group. The smaller island of Flores, with its adjacent islet of Corvo, will make the most northerly and westerly group. The five remaining islands, being clustered near together and between the two outlying pairs, make up the main or middle group. So far as hitherto ascertained apart, the floras of the three groups may be stated numerically thus :—

South-eastern 390. Middle 376. North-western 241.

The most numerous flora would thus appear to correspond with or belong to the group including the largest of the islands, and the one situate nearest to Madeira and the Continents. But the difference of numbers between this and the middle group is small, and may perhaps be sufficiently accounted for in the fact that its flora has been examined by a resident Botanist. The two islands least examined, San Jorge and Graciosa, being in the middle group, the flora of this group may be said virtually to be that of three islands, not of five. The north-westerly group has a much smaller surface of land, and its flora has been less examined than the floras of the other two groups; two circumstances tending to account for the much lower numerical amount of its flora as hitherto recorded.

The usual technical characteristic of a small flora is here very obvious,—that of great ordinal and generic diversity in proportion to the total number of species. The 478 species may be taken to represent 80 orders; being an average of only six species to an order. In the British islands, with a flora fully thrice as numerous in species, the orders are only about one-fourth more, while the average gives between fourteen and fifteen species to an order. The average for all Europe, as one whole, is 74 species to the order. In the Isles thirty-two of the orders are represented by single species; and about twenty (19) others have only two or three species each. It may be worth while to compare the rest of them (29) with an equal number in the flora of Britain, to illustrate the prevailing similarity, along with occasional differences, between the orders which predominate numerically in the two countries:—

In Azores.	In Britain.
1. Gramina, 51 species.	1. Compositæ, 135.
2. Compositæ, 43.	2. Gramina, 112.
3. Leguminiferæ, 43.	3. Cyperaceæ, 93.
4. Filices, 31.	4. Rosaceæ, 82.
5. Cyperaceæ, 22.	5. Leguminiferæ, 70.
6. Umbelliferæ, 20.	6. Cruciferæ, 60.
7. Cruciferæ, 19.	7. Umbelliferæ, 57.
8. Scrophulariaceæ, 17.	8. Caryophyllaceæ, 57.
9. Lamiaceæ, 16.	9. Scrophulariaceæ, 48.
10. Caryophyllaceæ, 13.	10. Lamiaceæ, 47.
11. Rosaceæ, 12.	11. Filices, 39.
12. Juncaceæ, 11.	12. Orchidaceæ, 39.
13. Ranunculaceæ, 9.	13. Ranunculaceæ, 35.
14. Polygonaceæ, 9.	14. Amentiferæ, 34.
15. Boraginaceæ, 8.	15. Polygonaceæ, 26.
16. Rubiaceæ, 7.	16. Juncaceæ, 26.
17. Euphorbiaceæ, 7.	17. Chenopodiaceæ, 24.
18. Chenopodiaceæ, 6.	18. Potamaceæ, 22.
19. Primulaceæ, 6.	19. Boraginaceæ, 21.
20. Geraniaceæ, 6.	20. Liliaceæ, 21.
21. Solanaceæ, 6.	21. Ericaceæ, 20.
22. Amarantaceæ, 6.	22. Rubiaceæ, 18.
23. Hypericaceæ, 5.	23. Primulaceæ, 17.
24. Plantaginaceæ, 5.	24. Geraniaceæ, 14.
25. Malvaceæ, 5.	25. Euphorbiaceæ, 14.
26. Lycopodiaceæ, 5.	26. Gentianaceæ, 14.
27. Potamaceæ, 4.	27. Saxifragaceæ, 14.
28. Gentianaceæ, 4.	28. Campanulaceæ, 14.

In Azores.	In Britain.
29. Papaveraceæ, 4.	29. Onagraceæ, 13.
30. Ericaceæ, etc. 3, or 4.	30. Crassulaceæ, 11.

In the above series, the numbers for orders in Britain are taken from the Cybele Britannica, where *Ericaceæ* and *Vacciniaceæ* are treated in combination as a single order. Hence the addition of "*Ericaceæ*, etc. 3, or 4" at the end of the column for the Azore orders; although in the present essay they are kept apart; and as orders respectively of 3 and 1 species they would not have found place in the above series.

Looking to the upper part of the series, the first eleven orders in the two lists are exactly the same; but *Filices* and *Rosaceæ* just change places, being fourth in the one list and eleventh in the other. The position of *Gramina* at the head of the list, and of *Compositæ* below them is a remarkable peculiarity in the sequence of orders in the Azoric flora. This numerical superiority of grasses and ferns may be attributed to the damp, mild, equable climate of the Isles. In accordance, we find *Juncaceæ* holding a high position also. But *Potamaceæ* and *Cyperaceæ* are relatively lower than in Britain; ponds and marshes being few, the streams rapid and rocky. Two orders which contribute much to the vegetal landscape in Britain, are subordinate in the Isles, *Rosaceæ* and *Amentiferæ*. In the former, we miss the genus *Rosa*, and the species of *Rubus* are taken at two only; the suborder *Pomaceæ* being wholly absent; that of *Drupaceæ* being reduced to a single species, the local *Prunus lusitanica*, an evergreen not adapted to remind the English traveller of the cherries and blackthorns of his own country. *Amentiferæ* also are reduced to a single representative, the evergreen *Myrica* or *Faya*, which passes for an "*Arbutus*" with non-botanical visitors in the Isles, and the general aspect of which is *ericaceous* rather than *amentaceous*. The most numerous order in the flora of Britain, which is entirely wanting in the flora of the Isles, is that of *Saxifragaceæ*; one which has small effect in the vegetal landscape of Britain, unless on the northern mountains. The more southern and lowland order *Orobanchaceæ* comes next; and its non-appearance in the Isles would scarcely be noticed except by a statistical botanist.

But the species-arithmetic of the orders is inadequate to show the conspicuous characteristics of the Azore flora and vegetation. *Verdure* is the chief feature or peculiarity of the Isles; and it may have been much more so before their surface was brought under cultivation by mankind. Evergreen shrubs and small trees, with ferns and mosses, must once have given the principal characteristics of the vegetation. And these may be said still to do so, along with grasses and other constituents of the turf or greensward, wherever the ground is left uncultivated, or the natural covering of shrubbery wood is allowed to remain or to renew itself. The sub-arborescent and frutescent species of *Juniperus, Erica, Laurus, Faya, Vaccinium, Myrsine, Hypericum,*—still abundant, and of *Myrtus, Prunus, Viburnum, Ilex, Daphne, Hedera, Persea, Picconia,*—now local or less plentiful, must have formerly covered the ground with a close forest of evergreens; unless, indeed, the two last are incorrectly placed among the aboriginal trees of the Isles. Deciduous shrubs are very few. *Rubus Hochstetterorum*, a very large bramble, is probably a subevergreen in these Isles, retaining its summer leaves into or through their soft winter. The planted orchards of orange and lemon trees correspond with the native flora in their evergreen foliage. Apparently, the herbaceous plants peculiar to the Isles on present knowledge, are mostly evergreen perennials, or biennials which grow through the winter to flower the succeeding spring or summer. But this statement can be made only as a probability by a writer who has not himself wintered in the Isles.

Regarded as a whole, the flora of the Isles corresponds closely with that of South-western Europe, as before remarked; although it is only a fragment of the far more numerous flora (say, for instance) of the Spanish peninsula, the nearest European land. So far as hitherto ascertained, the following species or segregates seem to be absolutely restricted to the Isles; though it may appear not unlikely that some of these will eventually be found elsewhere:—

Cardamine Caldeirarum.
Nasturtium flexuosum.
Cerastium azoricum.
Hypericum foliosum.
Vicia Dennesiana.
Rubus Hochstetterorum.
Sanicula azorica.
Ammi Huntii.
Petroselinum trifoliatum.
,, Seubertianum.
Solidago azorica.
Seubertia azorica.
Senecio malvæfolius.
Tolpis nobilis et varr.
Microderis rigens.
,, filii.
Campanula Vidalii.
Vaccinium cylindraceum.
Erica azorica.
Erythræa Massoni.
Veronica Dabneyi.
Euphrasia grandiflora.
Myosotis azorica.
,, maritima.
Lysimachia azorica.
Euphorbia azorica.
Habenaria micrantha.
,, longebracteata.
Luzula purpureo-splendens.
Carex azorica.
,, lævicaulis.
,, rigidifolia.
,, Hochstetteriana.
,, floresiana.

Carex Vulcani.	Holcus rigidus.
? Deyeuxia cæspitosa.	Festuca petræa.
? ,, azorica.	Isoetes azorica.

The forty species in the above list are all assigned to European genera with three exceptions. The two species of *Microderis* and one species of *Seubertia* belong to the vast order of *Compositæ,* in which generic divisions are made to rest on slight technical differences. *Microderis* might be held to form a section or subgenus of *Picris* or of *Crepis.* The *Seubertia* might easily pass for a *Bellis,* if seen in foliage only or even in the early flowering stage; but its rough and flattened receptacle, with the reflexed involucral scales, separate it from *Bellis* when in fruit.

The affinities of the forty species above mentioned are chiefly European, though not exclusively so. The species of *Hypericum, Senecio, Tolpis, Vaccinium, Luzula,* and *Festuca,* apparently have their nearest affinities in Madeira or the Canaries. Perhaps the affinity of *Ammi Huntii* is closer to the *Ammi procerum* of Madeira than to the *Ammi majus* of Europe. *Campanula Vidalii* diverges widely from all the European species, but without approximating towards the *Muschia* of Madeira or the *Canarina* of the Canaries. The *Erica, Erythræa, Lysimachia,* and *Isoetes* of the Isles, are so similar to European species that they have been published as simple varieties of them. Perhaps none of the others could be mistaken for species known in Europe, while they still bear so much resemblance to European plants that the discovery of them in Spain or Italy would not have seemed anywise extraordinary.

The subjoined list of plants found also in Madeira or the Canaries, one or both, but not in Europe unless by an accidental introduction, shows a partially exclusive affinity between the Azore flora and that of the neighbouring (but more African) islands of the Atlantic :—

Ranunculus grandifolius.	Laurus Canariensis.
Lepidium virginicum.	Persea indica.
Frankenia ericifolia.	Euphorbia mellifera.
Sida rhombifolia.	Myrica Faya.
Ilex Perado.	? Juniperus brevifolia.
Rhamnus latifolia.	Smilax canariensis
Pedrosia macrantha.	*Amaryllis Belladonna.
Aichryson villosum.	Carex sagittifera.
Torilis tenuifolia?	Deschampsia argentea.
? Hedera Canariensis.	Festuca jubata.
Bidens leucantha.	Dicksonia Culcita. (Spain?)
Tolpis fruticosa.	Asplenium monanthemum.
Thrincia nudicaulis.	Athyrium umbrosum.
Picconia excelsa.	Aspidium molle.
*Solanum Pseudo-capsicum.	Gymnogramme Lowei.
*Physalis pubescèns.	Acrostichum squamosum.
*Chenopodium ambrosioides.	Lycopodium suberectum.
Alternanthera Achyrantha.	Selaginella Kraussiana.

Short as this list is some exceptions might warrantably be taken against it. Four of the plants are marked (*) above as introductions into the Isles, and probably from more distant habitats. The *Bidens* and *Sida* may also be non-indigenous. The *Juniperus* and *Torilis* are dubiously distinct from kindred species of Europe. And it has been lately intimated that *Hedera canariensis* occurs in Portugal; but whether such is truly the fact, or the name has been misapplied, may be held uncertain for the present. The *Lycopodium* may perhaps be a variety of our *L. Selago*.

The list of non-European plants, wanting also in Madeira

and Canaries according to present incomplete accounts of their botany, is a very brief one :—

Lepidium virginicum. America; introduced to Madeira, etc.
Cakile americana. American coast.
Myrsine africana. Inter-tropical and South Africa.
Cyperus vegetus. America, with slight uncertainty.
Lycopodium cernuum. Intertropical America, etc.
Eleusine indica. Eastern. Casual in South Europe?

After deducting the 80 plants enumerated in these three lists, there will remain nearly 400 species common to Europe and the Azore Isles; including among them, however, a considerable number of plants doubtless introduced into the Isles from Portugal or other countries of Europe; also various others, concerning which there still remains some degree of uncertainty as to their nomenclature, or specific identity, etc. The subjoined numerical analysis will show passably well the floral affinities, some few probabilities reckoned in :—

Total Azore flora, taken at	480
Europe, more or less generally, about . .	400
Spain, Italy, France, each (severally) . .	360
Algeria 300. North Africa, say	320
Madeira 300. Canaries 260. Together .	340
Azore Isles exclusively, as yet known . .	40

The Spanish peninsula probably has more species in common with the Isles than either Italy or France; but its flora is yet incompletely on record. Certainly, there are several plants of the Isles which occur also in Spain or Portugal, and which apparently are not found elsewhere in Europe. The annexed list of some of these plants must be received with caution, as not being certainly correct in each instance, and it may be not complete.

Trifolium cernuum.	Vinca media.
Prunus lusitanica.	Corema alba.
?Hedera canariensis.	Trichomanes speciosum.
Rubia splendens.	Pteris arguta.
Anthemis aurea.	Asplenium palmatum.

We know that the *Trichomanes* occurs also in the British islands, though probably not elsewhere in Europe. Two other of our Ferns, the *Aspidium æmulum* and *Hymenophyllum unilaterale,* this latter extending northwards into Faroe and Norway, might well be expected in the Spanish peninsula, although Milde (1867) does not appear to have ascertained their existence there.

The flora of the Isles ought to bear importantly on the Forbesian hypothesis of a great Continental extension westward or south-westward from Europe in its present limits. How far will it give support or confirmation to that hitherto utterly unsubstantiated conjecture? The flora is European in its general character, and in detail it is numerously composed of European species; while nearly all of the additional species are found in the neighbouring island-groups of Madeira and Canary, or else are known only in the Azore Isles themselves. At first view, these facts may seem well in accordance with the hypothesis alluded to. And yet, in examining the more special details, they seem very difficult to reconcile with the idea of these Isles truly being the dissevered remnants of a great continental land formerly uniting them with Europe.

The plants which must be held specially to characterize the Azore flora, at the present time, are precisely those which seem least fitted to endure a continental climate; being unable to bear any extremes of heat and cold, and especially dryness of climate. When cultivated in England, a cold winter, a dry summer, are alike fatal to them. They succeed under glass in an unheated frame, if protected from frost in winter, and from dry heat in summer; the dryness apparently being more injurious to them than the summer heat of our own island. But there are considerably different adaptations or requirements of climate among them, as might be safely predicated of the plants of any country. For instance, *Myosotis azorica* will bear slight frost in winter, and it thrives well during that season in an unheated frame; but it is extremely impatient of sun and dryness in summer. On the contrary, *Campanula Vidalii* is apt to damp off in winter, in the frame which suits the *Mysotis*, while it bears the sun of summer very well. The *Cardamine caldeirarum, Cerastium azoricum, Vicia Dennesiana, Senecio malvæfolius, Erica azorica, Veronica Dabneyi*,—all resemble the *Myosotis* in their climatal requirements, more than they resemble the *Campanula*; their suitable climate being the unheated frame, protected from low temperatures in winter, from dry heat in summer. In short the present plants of the Azores, known only in those Isles, appear specially unadapted to endure a continental climate.

But there is another small group of the plants, having climatal requirements of similar character, which nevertheless do now occur wild in Europe, and some of which are familiar shrubs in the gardens of England; thus actually bearing a continental or sub-continental climate. Several *Ferns, Sibthorpia europæa, Umbilicus pendulinus, Rubia splendens, Menziesia polifolia, Viburnum Tinus*, and *Prunus lusitanica* are species common to the Isles and Western Europe, and not adapted to extreme climates. It may be thought possible that the other more restrictedly Azore plants might themselves have once existed on the oceanic outskirts of a great continent, or on islands immediately adjacent thereto; although now so exclusively insular in their actual habitat and in their climatal adaptations. Still, the possibility of them formerly bearing a continental or sub-continental climate so well as to hold their ground among other truly continental plants, is at best only a convenient guess; while their actual unsuitability to such a climate is a positive fact.

Another allied question may also arise in connexion with the Darwinian theories about the succession of life on our globe, and the mutation of its species. It has been explained that the affinity of the Azore flora is European in its general character, and that the affinity extends also to several of the species which are peculiarly Azoric; some few exceptions to this connecting the flora and species more closely with those of the Madeira and Canary group. Is the affinity sufficiently close and general to suggest a probability that the European and Azoric species could have diverged from the same species-ancestors in common since the severance of the Isles from Europe, supposing them to have been formerly united or nearly united therewith? An affirmative answer might be given in regard to some of them, hardly so in regard to all of them.

In those instances in which the affinity or resemblance is so close that doubts have actually arisen, and have been differently judged by different botanists, whether the plants ought to be held wide varieties or near species, it is at least as blindly bold to deny the possibility of such a divergence from common ancestors, as it would be to assert its probability. The shortened leaves and flowers, and the compact growth of the *Erica azorica*, might be held variations from the European *Erica scoparia*, in the direction which would better adapt the former to bear tempestuous winds, sweeping over the hilly surfaces of small islands; and we know that the climate of the Isles is changeable and tempestuous. The procumbent habit of *Lysimachia azorica*, in contrast against the more prostrate and rooting habit of its near European ally *L. nemorum*, may also better adapt the former to its situations of growth, on the sloping banks and steep rocks of the Isles; nor is it any serious claim on credulity, to suppose these two to be really divergent descendants from a common ancestor, more or less intermediate between them.

The equability of temperature and the mild dampness of the Azore Isles, are also conditions quite in natural correspondence with the evergreen and glabrous foliage of the *Veronica Dabneyi*, while the spreading wiry stems and the coriaceous texture of its leaves might also be held adaptations to its places of growth, on rocks much exposed to violent winds and subject to frequent showers. It is difficult in this case, however, to suggest any Europe *affine* or *analogue*, which could have diverged from the same single parental species; the divergence could only have been remotely ancestral. Still more difficult or impossible would it be, to name the European *Campanula* which could be accepted as a brother-species or cousin-species with the *Campanula Vidalii* of the Isles; although it is easy enough to look on this latter as a plant specially modified and adapted to its place of growth on coast rocks, under a mild and even temperature, but much subject to winter storms.

On the whole, while the peculiar plants of these Isles seem very well adapted to the physical and climatal characters of their actual habitat, they can hardly be said to yield any special evidence in support of the Darwinian theories. At the same time, it may be admitted, their affinities with the plants of Europe and Madeira, on the general view, are more in support of those theories than adverse to them.

It is otherwise with respect to the Forbesian hypothesis. The numerous identities between the species of Europe and the Azores might be said to give a general support to the hypothesis; but the special facts, the climatal requirements of the specially Azoric plants, are much adverse to that hypothesis.

The striking fact remains for explanation, as a yet unanswered query, Why have these small islands about forty species of plants not identified with anything found elsewhere, and almost as many more not found in Europe, but inhabiting Africa and its islands (chiefly) or America?

Some time after the foregoing pages were written, Mr. Godman requested me to add to them a List which would show the presence or apparent absence of each species in Europe (or Africa), Madeira, and the Canaries; and at the same time also distinguish those quite peculiar to the Isles, on present knowledge. As yet, there is no complete and perfected Flora for Madeira or for the Canaries, and some of the plants seemingly peculiar to the Azores, are in the doubtful category of species held to be rather local varieties than species absolutely distinct from those of Europe. The subjoined List must thus be received as approximate only.

List of Species.

	Europe.	Madeira.	Canaries.	America.	Isles only.	Africa.
1. Ranunculus Flammula	*	..	..	*		
2. —— grandifolius	..	*				
3. —— repens	*	*	..	*		
4. —— trilobus	*	*	*			
5. —— parviflorus	*	*	*	*		
6. —— muricatus	*	*	*	*		
7. Nigella arvensis	*					
8. Aquilegia vulgaris	*	*	*			
9. Delphinium Ajacis	*	*	*			
10. Papaver somniferum	*	*	*			
11. —— Rhœas	*	*	*			
12. —— dubium	*	*	*			
13. Chelidonium majus	*	*	*			
14. Fumaria muralis	*	*				
15. —— officinalis	*	..	*			
16. —— micrantha	*					
17. Matthiola annua	*					
18. Nasturtium officinale	*	*	*			
19. —— flexuosum	..	..	..	..	?	
20. Cardamine Caldeirarum	..	..	..	..	*	
21. —— hirsuta	*	*	..	*		
22. Barbarea præcox	*	*	*	*		
23. Sisymbrium officinale	*	*	*			
24. —— Irio	*	..	*			
25. Sinapis nigra	*	*				
26. Alyssum maritimum	*	*	*			
27. Senebiera pinnatifida	*	*	*			
28. —— Coronopus	*	*	*			
29. Lepidium virginicum	..	*	*	*		
30. Capsella Bursa-Pastoris	*	*	*			
31. Rapistrum perenne	*	*				
32. —— rugosum	*	*	*			
33. Cakile americana	..	..	..	*		
34. Raphanus Raphanistrum	*	*	*			
35. —— Landra	*					
36. Reseda Luteola, var.	*	*	*			
37. —— macrosperma	*					
38. Viola palustris	*	..	..	*		
39. —— odorata	*	*	*			
40. —— tricolor	*	*	..	*		
41. Polygala vulgaris	*					
42. Frankenia pulverulenta	*	*	*			
43. —— ericifolia	..	..	*			
44. Silene inflata? var.	*	*	*			
45. —— gallica	*	*	*			
46. Cerastium azoricum	..	..	..	..	*	
47. —— glomeratum	*	*	*			
48. —— triviale	*	*				
49. Stellaria media	*	*	*			

	Europe.	Madeira.	Canaries.	America.	Isles only.	Africa.
50. Mœhringia muscosa	*					
51. Sagina procumbens	*	*	*	*		
52. Spergula arvensis	*	*	*			
53. Spergularia rubra	*	*	*	*		
54. —— marina	*	..	*	*		
55. —— macrorhiza	*					
56. Polycarpon tetraphyllum	*	*	*			
57. Portulaca oleracea	*	*	*	*		
58. Elatine hexandra	*					
59. Hypericum foliosum	..	..	..	..	*	
60. —— perforatum	*	*	*			
61. —— bæticum	*	*	*			
62. —— humifusum	*	*				
63. —— elodes	*					
64. Malva parviflora	*	*	*			
65. —— nicæensis	*	*				
66. —— rotundifolia	*					
67. Lavatera sylvestris	*	*				
68. Sida rhombifolia	..	*	*	*		
69. Erodium malacoides	*	*	*			
70. —— moschatum	*	*	*			
71. Geranium rotundifolium	*	*	*			
72. —— molle	*	*	*			
73. —— dissectum	*	*	*			
74. —— Robertianum	*	*	*	*		
75. Oxalis corniculata	*	*	*			
76. Ruta bracteosa	*	*	*			
77. Ilex Perado	..	*	*			
78. Rhamnus latifolius	..	*				
79. Rhus Coriaria	*	*	*			
80. Sarothamnus scoparius	*	*				
81. Ononis arvensis	*					
82. Trigonella ornithopodioides	*	*				
83. Medicago lupulina	*	*				
84. —— lappacea	*	*	*			
85. —— denticulata	*					
86. Melilotus parviflora	*	*				
87. Trifolium angustifolium	*	*	*			
88. —— arvense	*	*	*			
89. —— ligusticum	*	*	*			
90. —— scabrum	*	*	*			
91. —— striatum	*	*	*			
92. —— maritimum	*	*				
93. —— lappaceum	*	*	*			
94. —— repens	*	*	..	*		
95. —— glomeratum	*	*	*			
96. —— suffocatum	*	*	*			
97. —— cernuum	*					
98. —— resupinatum	*	*	*			
99. —— subterraneum	*	*	*			
100. —— procumbens	*	*	*			

	Europe.	Madeira.	Canaries.	America.	Isles only.	Africa.
101. Trifolium minus	*	*				
102. Lotus major	*	*				
103. —— corniculatus	*					
104. —— angustissimus	*	*	*			
105. —— hispidus	*	*				
106. —— parviflorus	*	*				
107. —— creticus	*					
108. Pedrosia macrantha	..	*				
109. Arthrolobium ebracteatum	*	*	*			
110. Ornithopus perpusillus	*	*				
111. —— roseus	*					
112. —— compressus	*	*	*			
113. Vicia hirsuta	*	*	*			
114. —— gracilis	*	*	*			
115. —— Dennesiana	..	..	..	..	*	
116. —— albicans	*	*	*			
117. —— angustifolia	*	..	*			
118. —— bithynica	*					
119. Lathyrus Aphaca	*	*	*			
120. —— sativus	*	*	*			
121. —— Clymenum	*	*				
122. —— tingitanus	*	*	*			
123. Prunus lusitanica	*	*	*			
124. Spiræa Filipendula	*					
125. Rubus fruticosus	*	*	*			
126. —— Hochstetterorum	..	..	..	..	*	
127. Fragaria vesca	*	*	*	*		
128. Potentilla anserina	*	..	..	*		
129. —— reptans	*	*				
130. —— Tormentilla	*	*				
131. —— verna	*					
132. Alchemilla arvensis	*	*	*			
133. Agrimonia Eupatoria	*	*	*	*		
134. Poterium Sanguisorba	*					
135. Tillæa muscosa	*	*	*			
136. Umbilicus pedulinus	*	*	*			
137. Aichryson villosum	..	*	*			
138. Myriophyllum alterniflorum	*					
139. Callitriche verna	*	*				
140. Myrtus communis	*	*				
141. Peplis Portula	*					
142. Lythrum Hyssopifolia	*	*	*			
143. —— Græfferi	*	*	*			
144. Epilobium parviflorum	*	*	*			
145. Ecbalium Elaterium	*					
146. Sanicula azorica	..	..	..	..	*	
147. Conium maculatum	*	..	*			
148. Smyrnium Olusatrum	*	..	*			
149. Apium graveolens	*	*	*			
150. Helosciadium nodiflorum	*	*	*			
151. Ammi Visnaga	*	*	*			

	Europe.	Madeira.	Canaries.	America.	Isles only.	Africa.
152. Ammi Huntii	..	..	..	..	*	
153. —— majus	*	*	*			
154. Petroselinum trifoliatum	..	..	..	..	*	
155. —— Seubertianum	..	..	..	..	*	
156. —— sativum	*	*				
157. Pimpinella villosa	*					
158. —— dichotoma	*					
159. Chærophyllum aromaticum	*					
160. Fœniculum vulgare	*	*				
161. Crithmum maritimum	*	*	*			
162. Angelica montana	*					
163. Coriandrum sativum	*	*	*			
164. Daucus Carota	*	*				
165. Torilis tenuifolia	..	*				
166. Hedera canariensis	..	..	*			
167. Viburnum Tinus	*					
168. Sambucus nigra	*					
169. Rubia splendens	?					
170. Galium Mollugo	*					
171. —— palustre	*					
172. —— anglicum	*	*	*	*		
173. —— Aparine	*	*	*			
174. Aspera muralis	*	*	*			
175. Sherardia arvensis	*	*	*			
176. Valerianella dentata	*	*	*			
177. Scabiosa nitens	?					
178. Galactites tomentosa	*	*	*			
179. Cirsium lanceolatum	*					
180. Carduus pycnocephalus	*	..	*			
181. Centaurea melitensis	*	*	*			
182. Erigeron canadensis	*	*	..	*		
183. Conyza ambigua	*	*	*			
184. Solidago azorica	..	..	..	..	*	
185. Bellis perennis	*					
186. Seubertia azorica	..	..	..	..	*	
187. Chrysanthemum Myconis	*	*	*			
188. —— segetum	*	*				
189. —— coronarium	*	*				
190. Anthemis Cotula	*	*	*			
191. —— aurea	*	*				
192. Achillea Millefolium	*	*	..	*		
193. Gnaphalium luteo-album	*	*	*			
194. Filago germanica	*	..	*			
195. —— gallica	*	*	*			
196. Senecio vulgaris	*	*	*			
197. —— sylvaticus	*	*				
198. —— erraticus	*					
199. —— malvæfolius	..	..	..	..	*	
200. Bidens leucantha	..	*	*	*		
201. Calendula arvensis	*	*	*			
202. Xanthium Strumarium ?	*	*	*	?		

	Europe.	Madeira.	Canaries.	America.	Isles only.	Africa.
203. Xanthium spinosum	*					
204. Cichorium Intybus	*	*	*			
205. Tolpis nobilis	..	..	..	..	*	
206. —— fruticosa	..	*	*			
207. —— barbata	*	..	*			
208. —— umbellata	*	*				
209. Thrincia nudicaulis..........	..	*	*			
210. —— hirta..................	*					
211. Helminthia echioides	*	*	*			
212. Urospermum picroides	*	*	*			
213. Hypochœris glabra	*	*				
214. Taraxacum officinale	*	*	*			
215. Lactuca Scariola	*	*				
216. Sonchus oleraceus	*	*	*			
217. —— asper	*	*	*			
218. Crepis virens	*	..	*			
219. Microderis rigens	..	..	..	..	*	
220. —— filii	..	..	..	..	*	
221. Campanula Vidalii	..	..	..	..	*	
222. —— Erinus	*	*	*			
223. Vaccinium cylindraceum	..	..	..	..	*	
224. Calluna vulgaris	*					
225. Menziesia polifolia	*					
226. Erica azorica	..	..	..	..	*	
227. Picconia excelsa	..	*	*			
228. Vinca media.................	*					
229. Asclepias fruticosa	*					
230. Erythræa Centaurium........	*	..	*			
231. —— Massoni	..	..	..	..	*	
232. —— lutea.................	*	*	*			
233. Exacum filiforme............	*					
234. Convolvulus arvensis	*	*	*			
235. —— sepium...............	*	..	..	?		
236. Batatas littoralis	*					
237. Solanum nigrum	*	*	*			
238. —— villosum	*	*	*			
239. —— Pseudo-capsicum	..	*	*			
240. Physalis pubescens	..	*	..	*		
241. Hyoscyamus albus	*	*	*			
242. Datura Stramonium	*	*	*			
243. Acanthus mollis	*	..	*			
244. Verbascum virgatum	*					
245. —— spurium	*					
246. Veronica Anagallis	*	*	*	*		
247. —— Dabneyi	..	..	..	..	*	
248. —— officinalis..............	*	..	..	*		
249. —— serpyllifolia............	*	..	..	*		
250. —— arvensis	*	*	*			
251. Euphrasia grandiflora........	..	..	..	..	*	
252. Bartsia Trixago	*	..	*			
253. —— viscosa................	*					

	Europe.	Madeira.	Canaries.	America.	Isles only.	Africa.
254. Scrophularia Scorodonia	*	*	*			
255. —— Balbisii	*					
256. Antirrhinum Orontium	*	*	*			
257. Linaria Elatine	*	*	*			
258. —— spuria	*	..	*			
259. Sibthorpia europæa	*					
260. Digitalis purpurea	*	*				
261. Verbena officinalis	*	*	*			
262. Mentha rotundifolia	*	*	*			
263. —— sativa	*	..	*			
264. —— aquatica	*	*				
265. —— Pulegium	*	*	*			
266. Lycopus europæus	*					
267. Thymus angustifolius	*	*				
268. Origanum virens	*	*	*			
269. Calamintha officinalis	*	..	*			
270. —— Clinopodium	*	*				
271. Ballota nigra	*					
272. Lamium purpureum	*	*	*			
273. —— amplexicaule	*	*	*			
274. Stachys arvensis	*	..	*			
275. Nepeta Glechoma	*					
276. Marrubium vulgare	*	*	*			
277. Prunella vulgaris	*	*	*	*		
278. Myosotis azorica	..	..	..	..	*	
279. —— maritima	..	..	..	..	*	
280. —— arvensis	*	*				
281. —— stricta	*	..	..	?		
282. —— versicolor	*	..	*			
283. Cynoglossum pictum	*	*	*			
284. Heliotropium europæum	*	*	*			
285. Echium violaceum	*	*	*			
286. Myrsine africana	..	..	..	..	..	*
287. Lysimachia azorica	..	..	..	..	*	
288. Anagallis arvensis	*	..	*			
289. —— cærulea	*	..	*			
290. —— tenella	*					
291. Centunculus minimus	*					
292. Samolus Valerandi	*	..	*			
293. Plantago major	*	*	*			
294. —— lanceolata, var.	*	*				
295. —— Coronopus	*	*	*			
296. —— Serraria	*	..	*			
297. Littorella lacustris	*					
298. Statice Limonium?	*	..	..	*		
299. Phytolacca decandra	*	*	*	*		
300. Beta maritima	*	*	*			
301. Chenopodium murale	*	*	*			
302. —— ambrosioides	..	*	*	*	..	*
303. —— rubrum	*					
304. Atriplex Babingtonii	*					

	Europe.	Madeira.	Canaries.	America.	Isles only.	Africa.
305. Salsola Kali	*	..	..	*		
306. Amaranthus Blitum	*	*	*			
307. —— chlorostachys	*					
308. Euxolus deflexus	*					
309. Achyranthes argentea	*	*	*			
310. Alternanthera Achyrantha	*	..	*			
311. Illecebrum verticillatum	*	*				
312. Polygonum serrulatum	*	*	*			
313. —— maritimum	*	*	*	*		
314. —— aviculare	*	*	*			
315. Rumex aquaticus?	*					
316. —— crispus	*	*				
317. —— conglomeratus	*	*	*			
318. —— pulcher	*	*	*			
319. —— bucephalophorus	*	..	*			
320. —— Acetosella	*	*	*	?		
321. Daphne Laureola	*					
322. Laurus canariensis	..	*	*			
323. Persea indica	..	*	*			
324. Corema alba	*					
325. Euphorbia mellifera	..	*	*			
326. —— Lathyrus	*	..	*			
327. —— Peplis	*	..	*			
328. —— exigua	*	*	*			
329. —— Peplus	*	..	*			
330. —— azorica	..	..	..	..	*	
331. Mercurialis annua	*	*	*			
332. Urtica membranacea	*	*	*			
333. Parietaria officinalis	*	*				
334. —— lusitanica	*					
335. Ceratophyllum demersum	*					
336. Myrica Faya	..	*	*			
337. Juniperus brevifolia	?					
338. Serapias cordigera	*					
339. Habenaria micrantha	..	..	..	..	*	
340. —— longebracteata.	..	..	..	..	*	
341. Iris fœtidissima	*	..	*			
342. Trichonema Columnæ	*					
343. Amaryllis Belladonna	..	*	*	..	..	*
344. Allium Ampeloprasum	*	..	*			
345. —— subhirsutum	*					
346. Ruscus aculeatus	*	*				
347. Smilax canariensis	..	..	*			
348. Potamogeton pectinatus	*	..	..	*		
349. —— pusillus	*	*	*	*		
350. —— lucens	*	..	..	*		
351. —— polygonifolius	*	..	..	*		
352. Lemna minor	*	*	*	*		
353. Arum italicum	*					
354. —— Arisarum	*					
355. Alisma Plantago	*	..	..	*		

	Europe.	Madeira.	Canaries.	America.	Isles only.	Africa.
356. Luzula purpureo-splendens ..	..	..	..	..	*	
357. —— campestris	*	*	..	*		
358. Juncus effusus	*	*	*	*		
359. —— glaucus...............	*	*				
360. —— acutus	*	*	*			
361. —— maritimus	*	..	*	*		
362. —— capitatus	*	*	*			
363. —— tenuis	*	..	..	*		
364. —— bufonius	*	*	*	*		
365. —— supinus	*					
366. —— lampocarpus	*	*				
367. Cyperus longus	*	*				
368. —— esculentus	*	*				
369. —— vegetus...............	..	..	..	?		
370. Cladium Mariscus	*	..	*			
371. Scirpus maritimus	*	..	*	*		
372. —— setaceus	*					
373. —— Savii	*					
374. —— fluitans	*					
375. —— palustris	*	..	*	*		
376. —— multicaulis	*					
377. Carex sagittifera	..	*				
378. —— vulpina	*	..	*	*		
379. —— divulsa...............	*	*	*			
380. —— stellulata	*	..	..	*		
381. —— azorica...............	..	..	..	..	*	
382. —— flava................	*	..	..	*		
383. —— lævicaulis	..	..	..	..	*	
384. —— rigidifolia	..	..	..	..	*	
385. —— Hochstetterorum	..	..	..	..	*	
386. —— floresiana.............	..	..	..	..	*	
387. —— Vulcani	..	..	..	..	*	
388. —— pendula	*	*				
389. Anthoxanthum odoratum	*	..	*			
390. Panicum sanguinale	*	*	*			
391. —— Crus-galli	*	*	*			
392. Setaria glauca	*	*	*			
393. —— verticillata	*	..	*			
394. Cynodon Dactylon	*	*	*			
395. Eleusina indica	*	*	*	..	..	*
396. Arundo Donax	*					
397. Agrostis alba	*	*	..	?		
398. —— verticillata	*	..	..	?		
399. —— vulgaris ?	*	..	..	?		
400. Deyeuxia cæspitosa..........	..	..	..	..	*	
401. —— azorica...............	..	..	..	..	*	
402. Gastridium australe	*	*	*			
403. Polypogon monspeliensis	*	*				
404. —— maritimus	*					
405. Lagurus ovatus	*	*	*			
406. Piptatherum multiflorum	*	*				

	Europe.	Madeira.	Canaries.	America.	Isles only.	Africa.
407. Aira caryophyllea	*	*	*			
408. Deschampsia argentea........	..	*				
409. Avena hirsuta	*	..	*			
410. —— brevis	*					
411. Arrhenatherum avenaceum ..	*	*	*			
412. Holcus lanatus..............	*	*	*			
413. —— rigidus	..	..	..	..	*	
414. Kœleria phlæoides	*	..	*			
415. Poa annua.................	*	*	*			
416. —— trivialis	*	*	*			
417. —— Eragrostis	*	*	*			
418. —— rigida	*	*	*			
419. —— loliacea	*					
420. Briza maxima	*	*	*			
421. —— minor	*	*	*			
422. Triodia decumbens	*	*				
423. Cynosurus echinatus	*	*	*			
424. —— cristatus	*					
425. Testuca bromoides	*	*				
426. —— jubata	..	*				
427. —— petræa	..	..	..	..	*	
428. —— elatior	*					
429. Bromus madritensis	*	*	*			
430. —— rubens	*	*	*			
431. —— mollis, var.	*	*	*			
432. Hordeum murinum..........	*	*	*			
433. Triticum repens	*	*	*			
434. Brachypodium sylvaticum	*	*	*			
435. —— distachyon	*					
436. Lolium perenne	*	*				
437. —— multiflorum	*					
438. Gaudinia geminiflora	..	..	..	..	*	
439. Nardus stricta	*					
440. Dicksonia Culcita	..	*	*			
441. Hymenophyllum tunbridgense	*	*	*			
442. —— unilaterale	*	*	*			
443. Trichomanes speciosum	*	*	*			
444. Cystopteris fragilis	*	*	*	*		
445. Adiantum Capillus-Veneris ..	*	*	*			
446. Pteris arguta	*	*	*			
447. —— aquilina	*	*	*	*		
448. Lomaria Spicant	*	*	*			
449. Woodwardia radicans........	*	*	*			
450. Asplenium palmatum........	*	*	*			
451. —— anceps	*	*	*			
452. —— monanthemum	..	*	*	..	..	*
453. —— marinum	*	*	*			
454. —— lanceolatum...........	*	*				
455. —— Adiantum-nigrum	*	*	*			
456. Athyrium Filix-fœmina	*	*	..	*		
457. —— umbrosum	..	*	*			

	Europe.	Madeira.	Canaries.	America.	Isles only.	Africa.
458. Scolopendrium vulgare	*	*	*	*		
459. Aspidium angulare	*	*	*			
460. —— Filix-mas	*	*	..	*		
461. —— dilatatum	*	..	..	*		
462. —— æmulum	*	*				
463. —— molle	..	*	*	..	..	*
464. Polypodium vulgare	*	*	*	*		
465. Gymnogramme Lowei	..	*	..	..	..	*
466. —— leptophylla	*	*	*			
467. Acrostichum squamosum	..	*	..	..	..	*
468. Osmunda regalis	*					
469. Ophioglossum vulgatum......	*					
470. —— lusitanicum	*	*				
471. Lycopodium cernuum........	..	..	..	*	..	*
472. —— complanatum	*	..	..	*		
473. —— suberectum	..	*				
474. —— Selago	*	..	..	*		
475. Selaginella Kraussiana	..	*	..	..	..	*
476. Isoetes azorica	..	..	..	..	*	
477. Equisetum maximum	*	*				
478. —— incanum	*	*	*			

It was the wish of Mr. Godman that a column for America should be added in the preceding List. Accordingly this has been done, and a few marks have been inserted for *some* only of the plants supposed to be indigenous in America. But I have neither leisure nor inclination to fill in an American column properly. The subject is beset with difficulties and uncertainties of all sorts. Many of the ordinary weeds of cultivated ground and road sides in Europe, and similarly found in the Azore Isles, are now found also in North or South America or both; and yet they cannot properly be included among the real plants of the American flora. Again, where the nativity is less liable to doubt, the American forms are at any rate varieties in several instances; and it is often impossible to ascertain whether the European or Azorean form likewise occurs wild on the New Continent. Moreover, the facts to be made out are spread through so many works, and would require the sacrifice of so much time to collect and scrutinize, that it cannot be undertaken by myself, other pressing arrangements now preventing the attempt.

MUSCI.

By W. Mitten.

The total number of the species of Musci enumerated in the accompanying list as found in the islands of Madeira, the Canaries, and the Azores, amounts only to 151; of this number about ten are exclusively arboreal, or prefer to grow on decayed wood, eight are confined to rocks, one species is truly aquatic, and the remainder are terrestrial, or found occasionally in all situations. Ninety species are known to grow in the Canaries, nearly all of which were collected in Teneriffe; of these, 33 are peculiar to the Flora of the Atlantic Islands, and the remainder are common to Madeira, the Azores, and to Europe. The greatest number of these species appears to have been gathered by Despréaux, which, with the nearly coextensive collections made by Webb and Berthelot, were elaborated by Dr. Montagne in the 'Histoire Naturelle des Iles Canaries' par P. B. Webb et S. Berthelot, published in 1850; in this work 77 species are included. A few additional species were collected by Schmidt and are preserved in the Hookerian Herbarium; these were not known to Montagne at the time he prepared his account.

In Madeira, out of 96 species (nearly all collected by Mr. J. Y. Johnson, who visited the island for several years in succession) 28 species are peculiar; more recently Mandon has gathered the greater part of those before obtained, and in some instances very much more complete specimens, most kindly communicated to me by M. Bescherelle; besides these, a few others were procured by Mason. The most remarkable of the species gathered by Mr. Johnson were described in the Linnean Society's Journal, vol. viii.; and some of these were afterwards redescribed under different names by Juratzka, in the 'Botanische Zeitung,' January 19th, 1866.

In the Azores 44 species only have yet been gathered; of these, nine are peculiar. A few species were collected by Hunt, and given to Mr. Watson, who kindly presented the whole to me some time after his return from the islands; with one exception (*Hypnum Hochstetteri*, Schimp.), all the remainder have been collected by Mr. Godman.

Surveying the whole bryology of the Atlantic Islands, the most remarkable species are :—*Rhamphidium purpuratum*, which, although it has in some respects a similarity to certain European species belonging to other genera, has no known congeneric species nearer than the West Indies and S. America; and *Bartramia Webbii*, which, although closely allied to *B. brachypus*, Bruch et Schimper, and thought by Montagne to be identical, differing, as he supposed, only in the absence of a peristome, is yet different in the substance of its foliage.

Astrodontium canariense (a moss corresponding in most particulars with *Leucodon sciuroides*, but with a different habit), *Leptodon longisetus* (a remarkable moss, inasmuch as *L. Smithii*, found in the same island, is also found in remote regions, and without being elsewhere accompanied by any nearly allied species), and the three large mosses *Sciaromium spinosum*, *S. prolixum*, and *S. setigerum* are more nearly allied in structure as well as in appearance to the *S. hispidum* of New Zealand and Tasmania than they are to the smaller species inhabiting the streams of the Andes of Bogotá, there being no known allied species found in Europe, Africa, or North America.

Excepting some cosmopolite species, as *Funaria hygrometrica, Ceratodon purpureus,* and some others, the species which are found in Europe as well as in the Atlantic Islands are either those which are generally distributed, or are more abundant in those countries bordering on the Atlantic Ocean, and especially in the British Islands; and the peculiar species found in the Atlantic Islands have a particular interest for British bryologists; for as *Myurium hebridarum, Hookeria læte-virens, Orthotrichum vittatum, Bryum canariense,* and *Stereodon canariensis* have been found to belong to the British flora and have not yet been observed, or only, in the case of *Bryum canariense,* in a very few continental stations, it may be not improbable that some others of the species at present peculiar to the Atlantic Islands may be found to reach as far in their distribution as a few localities in the British Islands.

1. RHAMPHIDIUM, Mitt. Journ. Linn. Soc. (1868) vol. xii.

1. R. PURPURATUM, sp. n. Caulis erectus simplex, inferne radicellis ferrugineis tomentosus; folia a basi subquadrata, superne latiora, amplexantia, patentia, sensim subulato-angustata, canaliculata, apice obtusiusculo subintegerrimo, nervo percurrente, marginibus ad basin partis patentis sinuato-recurvis, cellulis in parte erecta angustis elongatis pellucidis, in parte superiore patente parvis ovalibus brevioribusque obscuriusculis; perichætialia conformia, parum longiora; theca in pedunculo purpureo oblongo-cylindracea inclinata subinæqualis lævis leptodermis, operculo anguste subulato recto æquilongo, peristomio dentibus ad basin divisis, laciniis rubris punctulatis angustis.

Azores, St. Michael (*Godman*); Madeira (*Johnson*).

Stems in the Azorean specimens densely tufted, an inch high, the lower half matted together with rusty rootlets; the upper portion dull yellowish green, tinged with brown, scarcely altered when dry. Seta scarcely half an inch high. Capsule of a red-brown colour and thin papery substance, about a line long, at base tapering into the seta, inclined at an angle of 45°; the operculum is of the same colour as the capsule; and the peristome, which is about one-third of the length of the capsule, is composed of red teeth, which, after the fall of the operculum, remain in an erect position; the calyptra is pale brown, and reaches to the middle of the capsule.

The Madeiran specimens were only a few fragments picked from an *Anthoceros*, and merely suffice to indicate the presence of the same species in that island.

There is no European species yet known to which this moss can be well compared; in its foliage, when the stems are short, it resembles *Dicranella Schreberi*, to which the Madeiran specimens were supposed to belong; but in the more complete Azorean plants, the elongated stems and proportionally shorter leaves take away any near approach in appearance to that moss.

2. Ceratodon, Brid.

2. C. purpureus, Linn.

Madeira (*Johnson*).

3. Rhabdoweisia, Schimp.

3. R. curvipes (C. Müller, *Zygodon*).

Madeira (*Johnson, Mandon*).

This moss has been placed in the genus *Zygodon*; but, excepting in the absence of a peristome, it agrees more nearly with the *Rhabdoweisiæ*, in which should probably be included also the *Amphoridium lapponicum* (Hedw.) and the *A. Mougeotii*, Bruch et Schimp.

4. Dicranum, Hedw.

4. D. scottianum, Turn.

Canaries (*Webb et Berthelot*); Teneriffe (*Bourgeau*); Madeira (*Johnson, Mandon, Mason*); Azores (*Godman*).

Usually a little larger than British or French specimens.

5. Campylopus, Brid.

5. C. introflexus (Hedw.).

C. longipilus, Mont. Hist. Nat. des Iles Canar. p. 37.

C. polytrichoideus, De Not. Syll. Musc. Ital. p. 221.

Canaries (*Despréaux*); Teneriffe (*Bourgeau*); Madeira, Ribero de S. Jorge (*Mandon*); Fonte de Joao Perado (*Johnson*).

Variable in size and in the colour of its leaves; in some of the specimens the points of the leaves are reflexed and spreading, as is usual in those from the southern hemisphere.

6. C. fragilis (Dicks.).

Madeira, Fonte de Joao Perado (*Johnson*), Ribero Frio (*Mandon*).

7. C. azoricus, sp. n. Caulis humilis gracilis cæspitosus; folia caulina erecto-patentia, a basi ovato-lanceolata, sensim subulato-angustata, nervo basi latitudinis $\frac{1}{3}$ occupante supra folii medium totum folium constituente e cellulis firmis composito canaliculato apice subdenticulato, cellulis superioribus parvis quadratis trapezoideisque densis, inferioribus oblongis laxis pellucidis, alaribus magnis laxis fuscis subauriculari-impressis; comalia basi latiora, apice longius angustiusque attenuata ibique subdenticulato-scabra;

theca in pedunculo brevi flexo, oblonga æqualis plicata, operculo rostrato, peristomio dentibus rubris dicranis; calyptra fuscata, basi fimbriis sparsis brevibus subnuda.

Azores, St. Michael (*Godman*); Madeira, amongst *Pogonatum aloides* (*Johnson*).

Nearly resembling *C. torfaceus*, Bruch et Schimp., but a little more robust; all the specimens are more yellow than is usual in *C. fragilis*, and essentially differ in the firm substance of their nerves, and in the presence of distinct coloured alary cells.

6. Grimmia, Ehrh.

* *Eugrimmia.*

8. G. pulvinata (Linn.).
Azores, St. Michael (*Godman*).

9. G. trichophylla, Grev.
Madeira (*Johnson*). Specimens small, without fruit.

** *Guembelia.*

10. G. leucophæa, Grev.
Canaries (*Despréaux*), Teneriffe (*Bourgeau*).

*** *Dryptodon.*

11. G. acicularis (Linn.).
Madeira (*Johnson*). Fine and fertile specimens.

**** *Rhacomitrium.*

12. G. canescens, Dill.
Madeira (*Johnson*), without fruit.

13. G. lanuginosa, Dill.
Azores, St. Michael (*Godman*), with capsules.

7. GLYPHOMITRIUM, Brid.

14. G. NIGRICANS (Bruch et Schimp., *Ptychomitrium*).

Notarisia crispata, Mont. Hist. Nat. des Iles Canar. tom. iii. p. 41?

Canaries (*Despréaux*); Azores, St. Michael (*Godman*).

15. G. PULVINARE, Mitt. Journ. Proc. Linn. Soc. vol. viii.

Madeira (*Johnson*).

16. G. POLYPHYLLUM, Dicks.; (*Bryum*) *Ptychomitrium*, Bruch et Schimp.

Canaries and Madeira (*Webb*); Madeira (*Johnson, Mandon*); Azores, St. Michael (*Godman*).

The differences between the peristome of this species and *G. daviesii* are about the same as those existing amongst the *Grimmiæ*, if all the groups here included in that genus are estimated merely as sections.

8. LEUCOBRYUM, Hampe.

17. L. GLAUCUM (Linn.).

Canaries (*Webb*), Teneriffe (*Bourgeau, Leman*).

18. L. JUNIPEROIDEUM, Brid. i. p. 409.

Canaries, Teneriffe (*Bridel*); Madeira (*Johnson*); Azores (*Watson*).

This so closely resembles the preceding in appearance, that it may be easily overlooked; but it appears to have the lamina of the lower part of the leaf about twice as wide as it is in that species.

9. WEISSIA, Hedw.

* *Euweissia.*

19. W. CONTROVERSA, Hedw.

Canaries (*Bourgeau*); Madeira (*Johnson*); Azores (*Godman*).

** *Hymenostylium*, Brid.

20. W. RUPESTRIS (Brid.).

Gymnostomum stelligerum, Mont. Hist. Nat. des Iles Canar. tom. iii. p. 45.

Canaries (*Webb et Berthelot*).

21. W. REFLEXA (Brid.).

Madeira (*Johnson*).

Closely as this species resembles *W. tenuis* (*Gymnostomum tenue* of most authors), it seems to differ sufficiently in the form of its perichætia, which have a dilated or obovate clasping base, from which the upper portion of the leaf is much narrowed and reflexed.

22. W. CALCAREA (Nees et Hornsch., *Gymnostomum*).

Azores, St. Michael (*Godman*).

23. W. VERTICILLATA, Schw.

Canaries (*Webb et Berthelot, Despréaux*); Madeira (*Johnson, Mandon*).

The Madeiran specimens have narrower leaves than usual in European states of this species. The erosion of the margins of the leaf a short distance above the base, more evident in this species, is not absent in *W. rupestris*.

10. TORTULA, Hedw.

* *Trichostomum.*

24. T. BARBULOIDES, Brid.

T. barbula, Schw.

Canaries (*Webb et Berthelot*), Teneriffe (*Bourgeau*); Madeira (*Johnson*).

25. T. DIAPHANA, Brid. i. p. 577.

Canaries (*Rudley, Bory de St. Vincent*).

26. T. BRACHYDONTIA (Müll.).
T. mutabile, Bruch.
Canaries (*Webb et Berthelot*); Madeira (*Johnson*).

27. T. FLAVO-VIRENS (Bruch).
Azores, St. Michael (*Godman*).

28. T. CIRRIFOLIA, sp. n. Dioica; caulis humilis; folia a basi brevi erectiore parum latiore, utrinque ad margines cellulis oblongis rectangulis pellucidis areolata, exinde angustata linealia, apice nervo excurrente mucronata, erecto-patentia integerrima, canaliculata, cellulis superioribus rotundis obscuris; perichætialia basi latiora; theca in pedunculo elongato rubro, cylindracea, operculo subulato.

Madeira, on walls, Funchal (*Johnson*).

Leaves pale glaucous-green, becoming pale brown when old. In size and habit like *T. flavo-virens*, but with much narrower leaves; the hyaline cells at the base much less conspicuous.

29. T. SQUARROSA, De Notaris.
Canaries (*Despréaux*).
The fructification of this species is truly terminal, and only in appearance lateral from the growth of innovations.

30. T. REVOLUTA, Schrad.
Canaries (*Despréaux*).
Montagne includes under this name *T. hornschuchiana* as var. *β*, and does not distinctly say whether one or both were obtained by Despréaux.

** *Aloina.*

31. T. ALOIDES, Koch.
Canaries, Teneriffe (*Bourgeau*).
The specimen was labelled *Trichostomum barbula.*

*** *Barbula.*

32. T. FALLAX, Hedw.

Madeira (*Johnson*).

A few stems with fruit which may belong to this species, certainly not to *T. vinealis.*

**** *Desmatodon,* Brid.

33. T. TRUNCATA (Hedw.).

Pottia truncata, Schimp. Syn.

Canaries, Teneriffe (*Bourgeau*).

34. T. STARKEANA (Hedw.).

Weissia affinis, Hook. et Tayl.; Mont. Hist. Nat. des Iles Canar. tom. iii. p. 39.

Anacalypta starkeana, Schimp. Syn.

Canaries, Teneriffe (*Webb et Berthelot, Despréaux*).

35. T. MURALIS (Hedw.).

Canaries (*Despréaux*); Madeira (*Johnson*).

36. T. ACUMINATA, Sw.

T. marginata, Bruch et Schimp.

Madeira (*Johnson*); Azores, St. Michael (*Hunt, Godman*).

The Madeiran specimens present several forms, which, at first sight, might easily be mistaken for distinct species; but after many examinations it has been impossible to discover a reliable character by which they might be separated. One form has its leaves very short, ovate, or ovate-oblong, and the nerve produced considerably beyond the point; another has leaves almost ligulate, the nerve not excurrent into a point, but ending with the leaf. The Azorean specimens are more pellucid; but all agree in the more or less evident margination; and although the cells are more obscure in some of the specimens, there is no real difference from European examples, with which an authentic specimen from Swartz appears to exactly coincide.

37. T. CHLORONOTOS, Schultz.

Canaries (*Despréaux*), Teneriffe (*Schmidt*); Madeira (*Johnson*).

***** *Zygotrichia*, Brid.

38. T. CUNEIFOLIA (Dicks.).

Canaries (*Brid.* ii. p. 829).

This may have been one of the forms of *T. acuminata*.

39. T. SUBULATA (Linn.).

Canaries, Teneriffe (*Schmidt*).

****** *Syntrichia*, Brid.

40. T. LÆVIPILA (Brid.).

Canaries, Teneriffe (*Bourgeau*); Azores, St. Michael (*Godman*).

11. ANŒCTANGIUM, Schw.

41. A. COMPACTUM (Schleich.).

Teneriffe (*Schmidt*); Madeira (*Johnson*).

42. A. ANGUSTIFOLIUM, Mitt. Journ. Proc. Linn. Soc. vol. viii.

A. knyi, Jur. Bot. Zeit. 1866, p. 20.

Teneriffe (*Schmidt, Mann*); Madeira (*Johnson, Kny, Mandon*).

12. ORTHOTRICHUM, Hedw.

* *Orthophyllaria*.

43. O. DIAPHANUM, Schrad.

Canaries (*Despréaux*), Teneriffe (*Bourgeau*).

44. O. PUMILUM, Swartz.
Gomera (*Despréaux*).

45. O. TENELLUM, Bruch.
Madeira (*Mandon*, 15). (*O. paivanum*, Schimp.)
It is probable that it was this species which gave occasion for the insertion of the preceding by Dr. Montagne.

** *Ulota.*

46. O. CRISPUM, Hedw.
Canaries (*Despréaux*).

47. O. VITTATUM, Mitt. Journ. Proc. Linn. Soc. vol. viii. (*Ulota*).
Ulota calvescens, Wils. in Schimp. Bryol. Europ. Supp. t. 1.
Madeira, on trees in the mountains (*Johnson, Mandon*).

13. PHYSCOMITRIUM, Brid.

48. P. PYRIFORME, Linn. (*Bryum*).
Canaries (*Despréaux*); Azores, St. Mary (*Godman*).

14. ENTOSTHODON, Schw.

49. E. CURVISETUS, Schw.
Physcomitrium curvisetum, Brid.; Mont. Hist. Nat. des Iles Canar. tom. iii. p. 31.
Canaries (*Despréaux*).

50. E. TEMPLETONI, Schw.
Canaries, Teneriffe (*Schmidt*); Madeira (*Johnson, Mandon*); Azores, St. Mary (*Godman*); Fayal (*Watson*).

15. FUNARIA, Schreb.

* *Eufunaria.*

51. F. HYGROMETRICA, Hedw.
Canaries, Teneriffe (*Webb et Berthelot, Bourgeau*); Maleira (*Johnson*); Azores, St. Michael (*Godman*).

52. F. calvescens, Schw.
Canaries (*Webb et Berthelot*).

** *Plagiodus.*

53. F. fontanesii, Schw.
Canaries (ex *Herb. Montagne*), Teneriffe (*Bourgeau*); Madeira (*Johnson*).

16. Bartramia, Hedw.

* *Philonotis*, Brid.

54. B. rigida, De Notaris.
Canaries, Teneriffe (*Webb et Berthelot, Bourgeau*); Madeira (*Johnson, Mandon*); Azores, St. Michael (*Hunt, Watson*), Fayal (*Godman*).

55. B. fontana (Linn.).
Canaries, Teneriffe (*Bourgeau*); Madeira (*Johnson*). Specimens all without fruit.

** *Breutelia*, Schimp.

56. B. azorica, sp. n. Caules elongati subpinnati fasciculati subverticillatimve ramosi, inferne tomento ferrugineo obtecti; folia patentia a basi ovata plicata sensim angustata canaliculata, apice acutissima, nervo excurrente, margine inferne anguste reflexa, inde ad apicem usque serrulata, cellulis basi ad angulos paucis subquadratis, reliquis angustis papilliferis.

Azores, St. Michael (*Hunt, from Watson*), Fayal (*Godman*).

Stems about 3 inches high. Young foliage yellow, not shining.

Closely resembling *B. arcuata* in size and general appearance, but distinct in its leaves being patent from the very base, without the erect and clasping portion always present in that species. From *B. tomentosa*, Sw., with which, in the insertion of its leaves, it agrees, it differs in their more narrowly reflexed lower margins and less conspicuous cells in the angles at their base.

*** *Eubartramia.*

57. B. WEBBII, Mont. Hist. Nat. des Iles Canar. tom. iii. p. 28, t. 2. f. 2 (*Glyphocarpus*).

Canaries, Teneriffe (*Webb, Schmidt, Bourgeau*), Gomera (*Bourgeau*); Madeira (*Johnson*).

Great confusion exists respecting the positions of many of the species belonging to the natural groups which make together the genus *Bartramia*. *Glyphocarpa* of Brown (*Glyphocarpus*, Brid.) included originally two species (*G. capensis* and *G. quadrata*); and these are, so to speak, abortive forms of the section now known as *Breutelia*, having the thin leaves and narrow cells usual in that group; but *Glyphocarpus webbii* has densely areolated leaves, as in *Bartramia stricta*, and the capsule perfectly smooth, and not "insculpta."

58. B. STRICTA, Brid. ii. p. 45.

Canaries (*Webb et Berthelot*), Teneriffe (*Mann, Bourgeau*); Madeira (*Johnson, Mandon*).

17. BRYUM, Dill.

* *Ateleobryum.* Peristomium internum orbatum.

59. B. NOTARISII, Mitt. Journ. Proc. Linn. Soc. vol. viii. (excl. syn. *B. alpinum*, var. *gemmiparum*, De Notaris).

Mielichhoferia crassinervia, Jur. Bot. Zeit. 1866, p. 20.

Madeira (*Johnson, Kny*).

Other and much finer specimens, afterwards brought by

Mr. Johnson, have conclusively shown that this is distinct from the variety of *B. alpinum* above mentioned; but the first specimens closely resembled it in colour and general appearance. When in good condition the foliage is of a shining green or golden colour, becoming, when old, of a reddish brown; and the capsule, which is horizontal, is of the same colour as that of mature *Bryum atropurpureum.*

The description of *Mielichhoferia crassinervia* given by Juratzka appears to leave no doubt that the same moss is intended, although the fruit certainly arises from the axis of the stem, as in *Bryum*, and not from a lateral branch, as in *Mielichhoferia.*

** *Eubryum.*

60. B. ALPINUM, Linn.

Canaries (*Webb et Berthelot, Bourgeau,* B. turbinatum); Madeira (*Johnson*).

61. B. JULACEUM, Schrad.

Canaries, Teneriffe (*Schmidt*); Madeira (*Johnson*); Azores, St. Michael (*Godman*).

62. B. ARGENTEUM, Linn.

Canaries (*Despréaux*); Azores, St. Michael (*Godman*).

63. B. ATROPURPUREUM, Web. et Mohr.

Canaries (*Webb et Berthelot, Despréaux*); Madeira (*Mandon,* B. erythrocarpum; *Johnson*).

64. B. CANARIENSE, Brid.

Canaries, Teneriffe (*Rudley, Webb et Berthelot, Despréaux, Bourgeau*); Madeira (*Johnson, Bourgeau, Mandon*).

65. B. PSEUDOTRIQUETRUM, Hedw.

Madeira (*Johnson*).

66. B. CÆSPITICIUM, Linn.
Canaries (*Webb et Berthelot*).

67. B. CAPILLARE, Linn.
B. platyloma, Schw.; Mont. Hist. Nat. des Iles Canar. tom. iii.
Canaries (*Despréaux*), Teneriffe (*Mann*); Madeira (*Johnson*); Azores, St. Michael (*Godman*).
A specimen from Montagne, marked by him *B. platyloma*, does not appear to differ from the states of this species in which the border of the leaf is a little more evident.

68. B. OBOVATUM, Mitt. Proc. Linn. Soc. vol. viii.
Madeira (*Johnson*).

18. EPIPTERYGIUM, Lindb.

69. E. TOZERI, Grev. (*Bryum*).
Madeira, with fruit (*Johnson*); Canaries, on a specimen of *Anthoceras* (ex *herb. Montagne*); Azores, St. Michael (*Godman*).

19. MNIUM, Linn.

70. M. UNDULATUM, Hedw.
Teneriffe (*Despréaux*); Madeira (*Johnson*), with fruit (*Mason*); Azores, Flores (*Godman*).

71. M. AFFINE, Bland.
Madeira, barren (*Johnson*); Teneriffe, with male flowers (*Mann*).

72. M. ROSTRATUM, Schrad.
Madeira, barren (*Johnson*).

20. Leucodon, Schw.

73. L. sciuroides (Linn.).
Canaries (*Despréaux*); Madeira (*Mandon*).

21. Astrodontium, Schw.

74. A. canariense, Schw.
Canaries (*Webb et Berthelot, Despréaux*), Teneriffe (*Schmidt, Bourgeau*); Madeira (*Johnson, Mandon*); Azores St. Michael (*Godman*).

22. Antitrichia, Brid.

75. A. curtipendula, Linn.
Canaries (*Webb et Berthelot*); Madeira (*Johnson*).

23. Cryphæa, Mohr.

76. C. heteromalla (Dill.).
Canaries (*Despréaux*).

24. Leptodon, Mohr.

77. L. smithii (Dicks).
Canaries, Teneriffe (*Despréaux*).

78. L. longisetus, Mont. Hist. Nat. des Iles Canar. tom. iii. p. 20.
Canaries, Teneriffe (*Webb et Berthelot, Bourgeau*).

25. Hookeria, Sm.

* *Euhookeria.*

79. H. lucens (Linn.).
Madeira (*Mandon*).

** *Cyclodictyon.*

80. H. LÆTE-VIRENS, Hook. et Tayl.
Madeira (*Johnson*).

26. LEPIDOPILUM, Brid.

Sect. *Tetrastichium.*

81. L. FONTANUM, Mitt. Proc. Linn. Soc. vol. viii.
Madeira (*Johnson*) ; Azores, Flores (*Godman*).
When the fructification of this curious species shall have been discovered, it will probably constitute a genus distinct from the S. American species it approaches in its foliage.

27. HEDWIGIA, Ehrh.

82. H. CILIATA (Dicks.).
Madeira (*Johnson*) ; Canaries, Teneriffe (*Mann*).

28. NECKERA, Hedw.

83. N. CRISPA (Linn.).
Canaries, Teneriffe, on laurels chiefly (*Webb et Berthelot*), on rocks (*Despréaux*) ; Madeira (*Johnson*).

84. N. INTERMEDIA, Brid. ii. p. 241.
N. elegans, Jur. Bot. Zeit. 1866, p. 20.
Canaries, Teneriffe (*Rudley, Bourgeau*); Madeira (*Johnson, Mandon, Milne, Kny*).

85. N. COMPLANATA, Linn.
Canaries, Gomera (*Despréaux*).

86. N. PUMILA, Hedw.
Canaries (*Despréaux*).

87. N. PENNATA, Hedw.
Canaries, Gomera (*Despréaux*).

29. Homalia, Brid.

88. H. webbiana, Mont. Hist. Nat. des Iles Canar. tom. iii. p. 12, t. 1. f. 1.

Canaries, Teneriffe (*Webb, Despréaux*).

89. H. subrecta, Mitt. Journ. Proc. Linn. Soc. vol. viii. Madeira (*Johnson*).

30. Thamnium, Schimp.

90. T. alopecurum (Linn.).

Canaries (*Webb et Berthelot, Despréaux*), Teneriffe (*Bourgeau*); Madeira (*Johnson, Mandon*).

31. Fontinalis, Dill.

91. F. antipyretica, Linn.

Madeira (*Johnson*); Azores, Flores (*Godman*).

32. Sematophyllum, Mitt. Journ. Proc. Linn. Soc. vol. viii.

92. S. auricomum, Mitt. *l. c.*

Madeira (*Johnson, Mandon*, no. 45); Canaries, Teneriffe (*Bourgeau*, no. 617, *Isothecium crassiusculum*).

33. Myurium, Schimp.

93. M. hebridarum, Schimp.

Madeira (*Johnson*); Azores (*Hunt*).

To this may belong the *Neckera imbricata*, Schw., mentioned in Martius, Fl. Brasil. 1.

34. Pterogonium, Sw.

94. P. gracile, Hedw.

Canaries (*Webb et Berthelot, Despréaux, Bourgeau*); Madeira (*Johnson*).

35. Pterygynandrum, Hedw.

95. P. filiforme, Hedw.
Canaries (*Despréaux*).

36. Ctenidium, Schimp., cum *Hylocomio* ejusd.

96. C. berthelotianum, Mont. Hist. Nat. des Iles Canar. tom. iii. p. 4, t. 1. f. 2 (*Hypnum*).

Teneriffe (*Webb*); Madeira (*Johnson*); Azores, St. Michael (*Hunt, Godman*).

This fine moss varies greatly in its appearance, the branches being sometimes pinnate and slender, sometimes less branched and much thicker and the leaves more closely imbricated; in this state it simulates very closely *Myurium hebridarum.*

To this species is referred, by C. Müller, the *Hypnum hochstetteri,* Schimp., in Seubert's 'Fl. Azorica;' but when *Myurium hebridarum* was sent to Dr. Schimper, and he was asked if that moss, being then known to have been gathered in the Azores and Madeira, was his *H. hochstetteri,* he did not notice the inquiry; and in Mandon's collection *C. berthelotianum* is labelled *Hylocomium maderense,* Schimp.; so that this reference may be questionable.

37. Plagiothecium, Bruch et Schimp.

97. P. sylvaticum (Linn.).
Madeira (*Johnson*).

38. Stereodon, Brid., ex parte.

98. S. cupressiforme (Linn.).

Canaries (*Webb et Berthelot, Bory, Despréaux*), Teneriffe (*Bourgeau*); Madeira (*Johnson*); Azores, Fayal (*Godman*).

Two states of this species are mentioned by Bridel, the

var. *longisetum* and the var. *lauri* (Brid. ii. pp. 609 et 611).

Mr. Johnson gathered the var. *compressum* and some other states not different from those common in Britain.

99. S. CANARIENSE, Mitt. Journ. Proc. Linn. Soc. vol. viii.
Hypnum uncinulatum, Jur. Bot. Zeit. 1866, p. 21?
Madeira (*Johnson, Mandon, Kny*); Canaries, Teneriffe (*Bourgeau*); Azores, St. Michael (*Godman*).

As this species is known to grow in Ireland, it may be expected to occur in some parts of Spain or Portugal.

39. PLEUROZIUM, Sullivant.

100. P. SPLENDENS (Hedw.).
Canaries, Gomera (*Despréaux*).

40. HYLOCOMIUM, Schimp.

101. H. SQUARROSUM (Linn.).
Azores, Fayal (*Godman*), sterile.

41. FABRONIA, Raddi.

102. F. PUSILLA, Raddi.
Madeira (*Johnson*).

42. LESCUREA, Schimp.

103. L. STRIATA, Schw.
Canaries (*Despréaux*).

43. HYPNUM, Dill.

* *Rhynchostegium*, Schimp.

104. H. RUSCIFORME, Weiss.
Canaries (*Webb et Berthelot*).

105. H. FONTIUM, Brid. ii. p. 417.
Canaries, Teneriffe (*Bory de St. Vincent*).

106. H. MEGAPOLITANUM, Bland.
Canaries (*Webb et Berthelot*).

107. H. CONFERTUM, Dicks.
Canaries (*Webb et Berthelot*); Madeira (*Johnson, Mandon*); Azores, St. Mary (*Godman*).

108. H. SURRECTUM, Mitt. Journ. Proc. Linn. Soc. vol. viii.
Madeira, on stones (*Johnson*).

109. H. TENELLUM, Dicks.
Madeira (*Johnson*).

110. H. BOURGEANUM, sp. nov. Monoicum; caulis depressus intertextus ramosus; folia patentia laxiuscule inserta elliptico-oblonga acumine subulato terminata parum concava, nervo indistincto medio evanescente, margine subserrulata, cellulis elongatis obscuriusculis, basalibus ad angulos abbreviatis subquadratis; perichætialia erecta e basi late ovata subulato-attenuata, subintegerrima, nervo valde indistincto; theca in pedunculo elongato rubro lævi, oblonga horizontalis, operculo subulato.

Canaries, Teneriffe (*Bourgeau*, no. 1235, *H. teneriffæ*).

This agrees in appearance so nearly with *H. teneriffæ* that it may easily be overlooked; but in its smooth seta it is more nearly allied to *H. tenellum*, yet different from both these species in the form of the leaf.

111. H. TEESDALII, Sm.
H. teneriffæ, Mont. Hist. Nat. des Iles Canar. tom. iii. p. 3, t. 3. f. 1.
Canaries, Teneriffe (*Webb et Berthelot*); Madeira (*Johnson*).

** *Eurhynchium*, Schimp.

112. H. PUMILUM, Wils.
Madeira (*Johnson*).

113. H. SWARTZII, Turn.
Madeira (*Johnson*); Azores, St. Mary (*Godman*).

114. H. PRÆLONGUM, Dill.
H. stokesii, Turner.
Madeira (*Johnson*); Canaries, Teneriffe (*Bourgeau*); Azores, St. Mary (*Godman*).

115. H. LONGIROSTRE, var. DURIÆI, Mont.
Madeira (*Johnson*); Azores (*Godman*).

*** *Scleropodium*, Schimp.

116. H. ILLECEBRA, Linn.
Canaries (*Despréaux*), Teneriffe (*Bourgeau*); Madeira (*Johnson*); Azores (*Godman*).

In some of the specimens the foliage on the principal branches is strongly plicate; and if it were not evident from the presence of branches arising from them with leaves quite smooth as is usual in British states of the species, they would scarcely be recognized as belonging to the same moss.

**** *Argyrodinum*.

117. H. PURUM, L.
Azores, St. Mary (*Godman*).

This elegant moss, although resembling *H. Schreberi*, does not well agree with any yet constituted group.

***** *Pleuropus*, Griff. *Homalothecium*, Schimp.

118. H. SERICEUM, Linn.
Canaries, Teneriffe (*Webb et Berthelot*); Madeira (*Johnson, Mandon*).

119. H. MANDONI, n. sp. Folia ramea ovato-lanceolata acuta plicata serrulata; perichætialia elongata sublævia, in acumen subulatum angustum educta; theca oblonga suberecta subinæqualis, operculo conico curvato.

Madeira, Ribiera das Calas (*Mandon*, 36). (*Homalothecium sericeum.*)

This closely resembles the common states of *H. sericeum* in size and habit; but its foliage is more closely imbricated and the points of the leaves are not produced into a narrow point, but are simply acute, so that, compared side by side with leaves taken from a corresponding part of *H. sericeum*, the difference in their respective outline appears so great that it seems almost impossible that they can belong to the same species; further investigation may reveal other discrepancies, or some intermediate states may be found. The capsule is not tapered upwards; but the peristome does not seem to be different from that of *H. sericeum*.

120. H. LUTESCENS, Hedw.
Canaries (*Despréaux*).

****** *Isothecium*, Brid., ex parte.

121. H. MYOSUROIDES, Linn.
Canaries (*Webb et Berthelot*), Teneriffe (*Bourgeau*).

******* *Brachythecium*, Schimp.

122. H. PLUMOSUM, Sw.
Madeira (*Johnson, Mandon*).

123. H. SALEBROSUM, Hoffm.
Canaries (*Webb et Berthelot*).

124. H. RIVULARE, Bruch et Schimp.
Madeira (*Mandon, Johnson*).

44. Amblystegium, Schimp.

* *Euamblystegium.*

125. A. maderense, Mitt. Journ. Linn. Soc. vol. viii. (*Hypnum*).

Madeira (*Johnson, Mandon*).

126. A. varium, Beauv.

Madeira (*Johnson*).

A small barren specimen, but not differing from the *A. radicale* of Bryol. Europ. *Hypnum radicale,* Beauv., according to an authentic specimen in Herb. Hooker, is certainly distinct from *H. varium,* and is in fact very closely similar to *H. serpens;* its leaves areolated with longer narrow cells; it appears to be common in British North America, and more rare in Europe. *A. varium* would appear to be a species almost confined to stagnant water in the south of England; but *A. serpens* is common by the sides of every rill, as well as in a great variety of situations remote from water.

127. A. riparium (Linn.).

Canaries, Teneriffe (*Webb et Berthelot*); Madeira (*Johnson*).

** *Harpidium,* Sull.

128. A. fluitans, Dill.

Canaries (*ex Montagne*).

*** *Acroceratium.*

129. A. cuspidatum, Linn.

Azores, St. Mary (*Godman*).

This species, together with *Hypnum giganteum,* Schimp., *H. cordifolium,* Hedw., *H. Richardsoni,* Mitt., and *H. sarmentosum,* Wahl., form a small group corresponding very nearly in habit and structure; all have the leaves at the apices of the stems imbricated into a smooth point; and they do not harmonize well in habit with the group which has been named *Harpidium,* nor with *Amblystegium,* although the capsules are alike in all these sections.

45. Sciaromium.

Sciaromium, Mitt. Journ. Proc. Linn. Soc. vol. viii., sub *Leskea.*

Scleromnion et *Echinodium,* Juratzka in Bot. Zeit. 1866, pp. 20, 21.

130. S. spinosum, Mitt. *l. c.* t. 1.
Madeira (*Johnson, Mason*); Azores (*Hunt*).

131. S. prolixum, Mitt. *l. c.*
Scleromnion Knyi, Jur. Bot. Zeit. 1866, p. 21.
Madeira (*Johnson, Kny*); Azores, Fayal (*Godman*).
The Azorean specimens have the extremities of their branches attenuated and branched in the same manner as in *Isothecium myosuroides.*

132. S. setigerum, Mitt. *l. c.*
Echinodium maderense, Jur. Bot. Zeit. 1866, p. 21.
Madeira (*Johnson, Kny*).

46. Thuidium, Schimp.

133. T. tamariscinum (Hedw.).
Madeira (*Johnson*); Azores, St. Michael (*Godman*).

134. T. minutulum (Hedw.).
Madeira (*Johnson*).

47. Fissidens, Hedw.

135. F. serrulatus, Brid. ii. p. 704.
Teneriffe (*Bory de St. Vincent, Webb et Berthelot, Bourgeau*); Madeira (*Johnson*); Azores, St. Mary (*Godman*).

This large and handsome species is now known to grow in Portugal and in Sardinia, whence specimens intermixed with *F. adiantoides* were sent by Serrafino to Gay. Specimens are also said to have been collected in Ireland; but by whom, is unknown.

136. F. ASPLENIOIDES, Swartz.

F. flabellatus, Hornsch.; Endlich. et Martius, Fl. Bras. i. t. 2. f. 2.

Madeira (*Johnson*); Canaries, Teneriffe (*Martius*); Azores, St. Michael (*Godman*).

137. F. TAXIFOLIUS (Linn.).

Canaries, Teneriffe (*Bourgeau*).

138. F. PALLIDICAULIS, sp. n. Monoicus; caulis brevis vel elongatus, interdum ramosus, sæpe subalbidus; folia approximata linealia, apice acuta, lamina vera ad medium producta, apice subæquali, lamina dorsali in caule latiuscule decurrente; omnes laminæ margine minutissime crenulatæ, concolores, nervo concolori vel flavescente percurrente, cellulis minutis rotundis glauco-viridibus obscuris; perichætia basilaria foliis paucis apicibus angustis elongatis; theca in pedunculo rubro, oblonga, horizontalis, operculo rostrato.

Madeira (*Johnson, Mandon*); Canaries, Teneriffe (*Bourgeau*, F. taxifolius).

Closely resembling *F. glaucescens*, Hornsch., from the Cape of Good Hope, and differing chiefly in the absence of a scarious or pellucid, as if thinner, margination of the lamina vera. From *F. taxifolius* it differs in the leaf being only half as wide, more narrow towards the point, with cells scarcely half so large, and the margin more minutely crenulate.

139. F. viridulus (Sw.).
Canaries, Teneriffe (*Bourgeau*); Madeira (*Johnson*).

48. Atrichum, Beauv.

140. A. undulatum (Linn.).
Madeira (*Mason*); Azores, Flores (*Godman*).

49. Pogonatum, Beauv.

141. P. urnigerum, Linn.
Canaries (*Despréaux*).

142. P. aloides, Hedw.
Canaries (*Despréaux*); Teneriffe (*Webb et Berthelot, Bourgeau*); Madeira (*Johnson, Mandon*).

143. P. nanum, Schreb.
Canaries (*Despréaux*); Teneriffe (*Bourgeau*); Madeira (*Johnson*).

50. Polytrichum, Dill.

144. P. piliferum, Schreb.
Canaries (*Webb et Berthelot*); Madeira (*Johnson*).

145. P. juniperinum, Willd.
Canaries (*Webb et Berthelot*), Teneriffe (*Bourgeau, Mann*); Madeira (*Johnson*).

146. P. commune, Linn.
Canaries (*Despréaux*); Madeira (*Johnson*); Azores (*Godman*).

147. P. formosum, Hedw.
Madeira (*Johnson*).

51. Diphyscium, Mohr.

148. D. foliosum, Linn.
Madeira (*Johnson*); a few barren plants only.

52. SPHAGNUM, Dill.

149. S. ACUTIFOLIUM, Ehrh.
Azores, Fayal (*Godman*).

150. S. CYMBIFOLIUM, Dill.
Azores, Fayal (*Godman*).

151. S. COMPACTUM, Brid.
Azores, St. Michael (*Godman*); Madeira (*Johnson*).

HEPATICÆ.

By W. MITTEN.

SIXTY-EIGHT species of *Hepaticæ* are known to occur in the Atlantic Islands; of these, forty-two are found in the Canaries, of which one only, and that probably doubtful, *Plagiochila javanica*, is peculiar, the remainder being common to the other islands, to the south of Europe, or North Africa. In Madeira thirty-eight species have been detected; of these, three are peculiar to that island. In the Azores sixteen species only can be as yet enumerated; of these, two are supposed to be peculiar, but both are somewhat doubtful. *Gymnomitrium erythrorhizum*, of which the description seems to agree so well with *Fossombronia angulosa* that it may be suspected an error similar to that which certainly occurred in the description of *Saccogyna viticulosa* as a new species of *Lophocolea* (*L. Preauxiana*) may have also happened in this instance. The other species peculiar to the Azores, *Rhacotheca azorica*, is possibly the same as *Fimbriaria africana*, which is found in the Canaries and Madeira.

Of the whole number of Atlantic-Island species, twenty-one are frondose, *i. e.* without distinct leaves; and it is almost certain, from some fragments brought from Madeira by Mr. Johnson, that there exists in that island another species; but these plants are revived with such difficulty after being once dried, that, unless the specimens are prepared with the utmost care whilst still living, their correct determination from the dried specimens is almost hopeless.

1. Gymnomitrium, Nees ab E.

G. erythrorhizum, Bischoff in Seubert et Hochstetter Fl. Azor.; Gottsche, Lindenb. et Nees, Syn. Hepat. p. 615.

Azores (*Hochstetter, jun.*).

From the description in the 'Synopsis Hepaticarum,' this must be far different from any other species referable to *Gymnomitrium*; but *Fossombronia angulosa* possesses so many of the characters ascribed to *G. erythrorhizum* in its size, foliage, and especially in the colour of its roots, as to lead to the inference that it was the species from which the description was drawn up.

2. Plagiochila, Mont. et Nees.

1. P. spinulosa, Dicks.

Gomera (*Despréaux*), Teneriffe (*Bourgeau*); Madeira (*Johnson*); Azores, Fayal (*Godman*).

Very variable in appearance, chiefly owing to different directions in which the leaves are disposed; for when removed from the stems their outline is not much varied.

2. P. javanica, Swartz.

Canaries (*Webb et Berthelot*).

It is most probable that some form of *P. spinulosa* was mistaken for this species.

3. LOPHOCOLEA, Nees ab E.

3. L. HETEROPHYLLA, Schrad.
Canaries (*Webb et Berthelot*).

4. L. BIDENTATA, Linn.
Canaries (*Despréaux*) ; Madeira (*Johnson*).
The Madeiran specimens are almost black, as remarked by Dr. Montagne to have been the case in those from the Canaries.

4. TRIGONANTHUS, Spruce.

5. T. BICUSPIDATUS (Linn.).
Azores, St. Michael (*Godman*).

6. T. CONNIVENS (Dicks.).
Madeira, creeping amongst *Sphagnum compactum* (*Johnson*).

7. T. DENTATUS (Raddi).
Jungermannia asperifolia, Tayl. Lond. Journ. of Bot. 1846, p. 277.
Madeira (*herb. Hooker*).
Stated to grow with *Campylopus clavatus*, by which was probably meant *C. introflexus*.

8. T. TURNERI (Hook.).
Canaries (*Webb et Berthelot*).

5. SARCOSCYPHUS, Corda.

9. S. EHRHARTI, Corda.
Canaries, Teneriffe (*Bourgeau*).

6. JUNGERMANNIA, Linn.

10. J. RIPARIA, Tayl.
Madeira (*Johnson*).

These specimens correspond entirely with British examples.

11. J. INFLATA, Hudson.
Canaries (*Webb et Berthelot*); Azores (*Godman*).

12. J. EXSECTA, Schmidel.
Madeira (*Johnson*).

13. J. ALBICANS, Linn.
Canaries (*Webb et Berthelot*); Madeira (*Johnson*).

7. SOLENOSTOMA, Mitt. Journ. Proc. Linn. Soc. vol. viii.

* *Alicularia*, Corda.

14. S. SCALARIS (Schrad.).
Teneriffe (*Bourgeau*, Jungermannia crenulata); Madeira (*Johnson*), intermixed with *Pogonatum aloides*.

The leaves on some of the shoots have frequently nearly all of them an obtuse sinus at their apices, which is more evident than is usual in British specimens.

15. S. HYALINA (Lyell).
Canaries (*Webb et Berthelot*); Madeira (*Johnson*); Azores, St. Michael (*Godman*).

All the specimens from the Atlantic Islands are a little more robust than European.

** *Eusolenostoma.*

16. S. CRENULATA (Smith).
Azores, St. Michael (*Godman*).

This species differs from the two preceding in having its perianth not adherent to the perichætial leaves. In *S. scalaris* both the upper involucral leaves are united to the sides of the perianth; in *S. hyalina* the perianth is united to one only, the other side being free.

All the species referable to this genus have the perianth, before the extrusion of the capsule, usually five-angled (or 5-plicate), the mouth contracted and ending in a small tube; and in this state the perianths are analogous to those of such species of *Lejeunia* and *Frullania* as have their perianths as many-folded; and all are alike burst along the keels of the folds by the egress of the capsule.

8. Chiloscyphus, Corda.

17. C. polyanthus (Linn.).
Madeira (*Johnson*).

18. C. denticulatus, sp. n. Caulis procumbens subsimplex; folia expansa, subovata, obtusa, apice denticulis 2–4 subintegerrimave, amphigastriis parvis bifidis laciniis utrinque unidentatis.
Madeira (*Johnson*).

Scarcely larger than *C. polyanthus,* but with foliage less altered by drying. Closely allied to *C. endlicherianus,* Nees ab E., from Norfolk Island, and to *C. argutus* from India and Java. The specimens retain a piperaceous odour after having been dried several years.

9. Saccogyna, Dumort.

19. S. viticulosa, Linn.

Lophocolea preauxiana, Mont. Hist. Nat. des Iles Canar. tom. iii. p. 50, t. 3. f. 3.

Teneriffe (*Webb et Berthelot*); Madeira (*Johnson*); Azores (*Godman*).

This species appears to be abundant in Madeira, being found intermixed with many mosses. Its having been mistaken by Montagne for a new species of *Lophocolea* may have arisen from his not being familiar with its usual appearance.

10. CALYPOGEIA, Raddi.

20. C. TRICHOMANIS, Spreng.

Madeira (*Johnson*); Azores (*Godman*).

These specimens do not appear to differ from small states found in the British Islands.

21. C. ARGUTA, Nees ab E.

Madeira (*Gottsche, Lindenb. et Nees, Synops. Hepat.* p. 199).

11. LEPIDOZIA, Nees, Lindenb. et Gottsche.

22. L. REPTANS (Linn.).

Madeira (*Johnson*).

12. NOWELLIA, gen. nov.

Caules prostrati. *Folia* succuba, directione incuba, margine inferiore (ventrali) inflexo saccato. *Amphigastria* caulina nulla. *Perianthium* in ramo brevi laterale, superne trigonum, apice truncatum.

23. N. CURVIFOLIA (Dicks.). Caules ramosi, radicellis paucis; folia transverse latiora, sursum secunda, patentia, dentibus elongatis duobus spiniformibus oblique incurvis, sinu intermedio lato, imbricata, lobulo subrotundo, apice angulato arcte appresso sacculum tumidum efformante; folia involucralia lobulis orbata, inferiora bidentata integerrima, superiora amphigastriumque coæquale bifida, lobis latis acutis serrulatis; perianthium elongatum, inferne teretiusculum, apicem versus triplicatum, ore denticulatum.

Jungermannia curvifolia, Dicks. fas. ii. t. 5. f. 7; Gottsche, Lindenb. et Nees, Synops. Hepat. p. 142.

Madeira (*Johnson*).

This very curious species appears to have been very imperfectly understood; for Dickson gives in his figure no idea of the real form of the leaves, and in Sir W. J. Hooker's 'Brit. Jungermanniæ' they are represented with the lobule on the upper instead of the inferior side of the leaf; hence the difficulty in exactly identifying specimens with the representations.

Without taking into consideration the perianth and its position, the stems and foliage of *N. curvifolia* have a much nearer resemblance to some tropical *Lejeuniæ* than to any species of *Jungermannia* or *Trigonanthus*. This genus is named in memory of the late John Nowell, of Todmorden, a zealous investigator of the Mosses and Hepaticæ of Yorkshire.

13. Scapania, Lindenb.

24. S. undulata (Linn.).
Canaries (*Despréaux*).

25. S. curta (Martius).
Canaries (*Webb et Berthelot*).

26. S. compacta (Roth).
Canaries, Teneriffe (*Bourgeau*); Madeira (*Johnson*).

27. S. nemorosa (Linn.).
Canaries, Teneriffe (*Bourgeau*); Madeira, abundant (*Johnson*).

14. Radula, Dumort.

28. R. complanata, Linn.
Canaries (*Webb et Berthelot*), Teneriffe (*Bourgeau*); Madeira (*Johnson*).

29. R. physoloba, Mont.
Madeira (*Johnson*); Azores, St. Michael (*Godman*).

15. Lejeunia, Libert.

30. L. serpyllifolia (Dicks.).

Canaries (*Webb et Berthelot*); Madeira (*Johnson*); Azores, St. Michael (*Godman*).

Several states of this occur in Madeira. One is rather larger than usual, and may be the following.

31. L. thymifolia, Nees ab E., var δ. *major*.

Madeira (*Ecklon*).

32. L. minutissima (Sm.).

Madeira (*Johnson*).

33. L. johnsoniana, Mitt. Journ. Proc. Linn. Soc. vol. viii. t. 2.

Madeira (*Johnson*).

This is unlike any European species, and resembles some tropical South-American forms.

34. L. hamatifolia (Hook.).

Madeira (*Johnson*), creeping over *Madotheca canariensis*.

16. Madotheca, Dumort.

35. M. lævigata, Dumort.

Canaries (*Webb et Berthelot, Despréaux*), Teneriffe (*Bourgeau*); Madeira (*Johnson*).

36. M. canariensis, Nees ab E.

Canaries and Madeira (*Webb*); Madeira (*Johnson, Mandon*).

37. M. platyphylla (Linn.).

Canaries (*vide* Syn. Hepat. p. 263).

17. Frullania, Raddi.

38. F. dilatata (Linn.).

Canaries (*Despréaux*).

39. F. HUTCHINSIÆ (Hook.).
Madeira (*Johnson*).

40. F. TAMARISCI (Linn.).
Canaries, Teneriffe (*Webb et Berthelot*).

41. F. HISPANICA, Nees ab E.
Canaries (*Webb et Berthelot*).

42. F. POLYSTICTA, Lindenb.
Madeira (*Holl.*, *Johnson*).

43. F. NERVOSA, Mont.
Canaries (Herb. *Montagne*).
This and the two preceding species may possibly be only states of *F. tamarisci*.

44. F. TENERIFFÆ, Weber.
Canaries (*Webb et Berthelot*), Teneriffe (*Bourgeau*); Madeira (*Johnson*); Azores (*Godman*).

18. FOSSOMBRONIA, Raddi.

45. F. PUSILLA (Linn.).
Canaries (*Despréaux*, the var. *capitata*, N. ab E.).

46. F. ANGULOSA, Raddi.
Teneriffe (*Bourgeau*); Madeira (*Johnson*); Azores (*Herb. Montagne*, *Godman*).

19. PELLIA, Raddi.

47. P. EPIPHYLLA (Linn.).
Azores, Flores (*Godman*).

20. ANEURA, Dumort.

48. A. MULTIFIDA (Linn.).
Azores, Flores (*Godman*).

21. Riccia, Micheli.

49. R. minima, Linn.
Canaries (*Despréaux*).

50. R. ciliata, Hoffm.
Canaries (*Despréaux*).

51. R. ciliifera, Link.
Canaries (*Despréaux*).

52. R. lamellosa, Raddi.
Canaries (*Despréaux*).

22. Corsinia, Raddi.

53. C. marchantioides, Raddi.
Canaries (*Despréaux*).

23. Targionia, Micheli.

54. T. hypophylla (Linn.).
Canaries (*Webb et Berthelot*); Madeira (*Johnson*).

24. Plagiochasma, Lehm. et Lindenb.

55. P. aitonia, Nees ab E.
Teneriffe (*Berthelot*).

25. Lunularia, Micheli.

56. L. vulgaris, Mich.
Madeira (*Johnson*); Azores (*Godman*).

26. Exormotheca, gen. nov.

Receptaculum femineum primo subglobosum, integrum, subtus pro insertione pedunculi perforatum, demum subturbinatum, loculis 1–4 protrusione capsulæ horizontali tubulosis, apice poro dilatato dehiscentibus. *Perianthium* nullum. *Calyptra* brevis, persistens, inclusa. *Capsula* dentibus irregularibus 4 dehiscens, in pedunculo brevi emergens.

57. E. PUSTULOSA, sp. n. Frondes cæspitosæ, dichotomæ, humidæ, superficie planiusculæ, siccæ autem marginibus sursum conniventibus complicatæ, sectione transversali trigonæ, costa nulla, subtus radicellis villosæ et ad margines squamis coloratis transverse oblongis obtusis integerrimis ultra marginem frondis exstantibus demum pallescentibus paleatæ; epidermis ubique papulis elevatis tenerrimis pallidis mamillis pustuliformibus, apice poriferis obtecta; stratum hypodermicum e cellulis oblongis, erectis, viridibus septatis formatum; reliqua pars frondis inferior e cellulis laxis hyalinis composita; pedunculus e sinu terminali vel inter dichotomiam e substantia frondis inferioris oriendus, basi nudus crassiusculus, carnosus, inferne striatus; receptaculum carnosum, apice papulis pustulosum, subtus pallidum, post egressu capsularum ad pedunculum contractum, filis paucis inconspicuis inclusis; capsula solitaria, per substantiam receptaculi protrusa lateraliter, ad horizontem erumpens, fusca, sporis tuberculosis fibrisque repleta.

Madeira, Pico de Barcellos (*Johnson*), with *Lunularia vulgaris*.

Fronds about half an inch long, nearly a line wide; the upper surface of a pale glaucous-green, from the translucency of the pustulous epidermis showing the green stratum beneath it; on each side they are bordered with projecting blackish-purple scales, which appear to become afterwards, by age, almost white. The peduncle is about half an inch long, of a pale brownish colour; it enters the receptacle by a small hole and is affixed to its internal substance near the upper surface. The receptacle appears to be at first nearly globose, but afterwards, by the pushing out of the capsules, the base becomes contracted and tapering to the peduncle; its substance is soft, and after having been dried it is almost impossible to get a complete idea of its proper form. The capsule appears to force its way through the sides of the receptacle in a horizontal or slightly ascending direction, leaving the orifice entire at its edge, without any trace of there having been any suture by which it had opened. No male organs or scyphi have been seen. This curious species differs from the *Lunulariæ* in its peduncle being naked at its base, and from the *Jecorariæ* in its capsules bursting out on the sides of the involucre, which is not radiate, but entire and contracted at its base.

27. MARCHANTIA, Raddi.

58. M. POLYMORPHA, Linn.
Canaries (*Webb, Despréaux*), Teneriffe (*Bourgeau*); Madeira (*Johnson*); Azores, St. Michael (*Godman*).

28. DUMORTIERA, Reinw.

59. D. IRRIGUA, Tayl.
Canaries, Palma (*Bourgeau*).

29. FEGATELLA, Raddi.

60. F. CONICA (Linn.).
Azores (*Watson*).

30. REBOULLIA, Raddi.

61. R. HEMISPHÆRICA, Linn.
Canaries (*Webb et Berthelot*).

31. Grimaldia, Raddi.

62. G. dichotoma, Raddi.
Teneriffe (*Webb et Berthelot, Bourgeau*).

32. Fimbriaria, Nees ab E.

63. F. africana, Mont.
Canaries (*Webb et Berthelot*), Teneriffe (*Bourgeau*); Madeira (*Johnson*).

33. Rhacotheca, Bischoff.

64. R. azorica, Bisch.
Azores (*Hochstetter*).
From the observations in the 'Synopsis Hepaticarum' of Gottsche, Lindenberg et Nees, it would seem that this is very near to *Fimbriaria africana*.

34. Anthoceros, Micheli.

65. A. punctatus (Linn.).
Canaries, the varieties *multifidus* and *crispulus* (*Webb et Berthelot, Despréaux*), Teneriffe (*Bourgeau*); Madeira (*Johnson*); Azores, St. Michael (*Godman*).

66. A. cæspiticius, De Notaris.
Canaries (*herb. Montagne*).

67. A. laciniatus, Schweinitz.
Madeira (*herb. Montagne*).

68. A. lævis, Linn.
Madeira (*herb. Montagne*).

SUMMARY.

ZOOLOGY.

MAMMALIA.

There appear to be no terrestrial Mammalia whatever indigenous to the islands. All the species we now find have been voluntarily or involuntarily introduced by human agency, *i. e.* either as associates of man or else for food-supply, or as followers in the track of man wherever he goes. Amongst the former are the Cat and Dog; Cattle, Goats, &c. may be mentioned as coming under the next heading; Rats and Mice, with the Weazel, under the last. Under these headings collectively the whole of the land Mammalia are included. One single species of Bat is found abundantly which is identical with a well-known European species, and whose presence in the Azores can easily be accounted for by accidental causes.

AVES.

The total number of Birds found in the islands is 53; but of these 15 must certainly be accounted accidental stragglers which have not yet gained a permanent footing in the islands. Of the 38 remaining, 18 or 20 only are to be accounted "land birds;" the remainder are either oceanic wanderers or frequent the sea-shore or river-banks. Three of the 20 strictly "land birds," or 15 per cent., are so far distinct from their continental allies that characters can be drawn out whereby they may be distinguished from the species most nearly related to them. These characters, it is true, are not trenchant; still they are constant, so far as I have observed, and must therefore be taken as indicating specific difference, if there is any meaning in the word. But though these three are the only species in which definable differences can be found, they by no means include all the species possessing, in a less marked degree, symptoms of divergence from their continental representatives. In certain other birds frequenting these islands, such as the Golden-crested Wren, Blackcap, and Rock-Dove, modifications are observable. These seem to be hardly sufficiently established as permanent characters to render their possessors separable as distinct species. The direction in which the modification seems universally to tend is towards the development of a more sombre cast of plumage, a greater strength of feet and legs, and a more robust bill. This tendency is not confined to the birds of the Azores, but is found in other island-faunæ. No more remarkable case than the singularly dull colouring of the birds of the Galapagos archipelago suggests itself.

REPTILIA.

There are no Reptiles or Snakes found in these islands, with the single exception of a Lizard (*L. dugesii*), whose presence in Graciosa is attributed by its discoverer, M. Drouet, to accidental introduction. At the same time, it must be remarked that, little communication existing between Graciosa and Madeira or the Canaries, I should certainly have expected that this species, if introduced, would be found in the more important islands.

AMPHIBIA.

Though now thoroughly established, *Rana esculenta* is certainly an introduced species. No other amphibian occurs.

PISCES.

Of the three species whose presence may now be noticed in the Azores, all are undoubtedly recent introductions.

COLEOPTERA.

The total number of species of Beetles is 212, of which one remains at present undetermined. These we may divide into European and non-European species. Of the former we find the number to be 175, whilst of the latter there are 36. Viewed in relation to the other Atlantic groups we see that of the European species 97 are also common to the other islands. Excluding the Canaries, we have 27 species common to the Azores, Europe, and Madeira; and excluding Madeira, we have 8 species common to the Azores, Europe, and Canaries. This leaves 43 species found in the Azores and Europe which are not found in the other Atlantic groups. Of the 36 non-European species, 8 are found in both the other groups, 8 in Madeira only, 3 in the Canaries only, 3 are South-American, leaving 14 species which have not yet been found outside the limits of the Azores.

HYMENOPTERA.

My materials are almost too scanty to give any general conclusions respecting the Hymenoptera of the Azores. The evidence, however, such as it is, shows that the European element largely preponderates. Some species, however, are found in Madeira and not elsewhere, whilst there are indications of the existence of species peculiar to these islands, as in the case of the Coleoptera.

LEPIDOPTERA.

All the Rhopalocera, with one exception, are European species. In this section of Lepidoptera we find no peculiar forms in any of the Atlantic Islands. There are, however, one or two instances of remarkable distribution, as shown by the occurrence of the American *Danais archippus* in the Azores and of *Vanessa hunteri* in Madeira, also an American insect.

My collection of Heterocera is very incomplete, numbering only some twenty species. By far the majority of these are European. One species appears to be restricted to the Azores and Madeira; but no peculiar species have as yet been discovered.

MOLLUSCA.

Considerable attention has been devoted to the terrestrial Mollusca of the Azores, resulting in the enumeration of 69 species. Of these, 26 are also found in Western Europe, 7 are common to Madeira, 4 only to the Canaries, leaving no less than 32 species peculiar to the Azores.

BOTANY.

Mr. Watson's list contains 478 species of plants, of which about 400 are found in the Azores and Europe, 340 are common to Madeira and the Canaries, 300 to Madeira, 260 to the Canaries, while 40 alone are peculiar. The Mosses have been rather differently treated by Mr. Mitten, whose paper gives a complete review of those known from all the Atlantic Islands. Their external distribution does not seem to have been specially considered. From the Azores he enumerates 44 species, 9 of which he considers peculiar; from Madeira 96 are given, 28 of which are peculiar; while from the Canaries 90 species are known to exist, of which 33 are peculiar to the Atlantic Islands. The remainder are possessed in common by the Atlantic Islands and Europe. Of Hepaticæ two (?) species out of 16 are peculiar to the Azores; 3 out of 38 peculiar to Madeira; and only one (doubtfully) out of 42 peculiar to the Canaries.

GENERAL REMARKS.

On taking a general survey of the natural productions of the Azores, the point about which the greatest amount of interest concentrates is the intricate question as to how the fauna and flora of these islands came to exist in the state we now see them. In endeavouring to find an answer, one becomes involved in some of the complicated problems which are now engrossing the attention of our leading naturalists. We find these isolated spots in the midst of a deep ocean. Were they always thus situated since the time when they first emerged from the water; or have they at some period long past formed a portion of a continent which extended far westward into the Atlantic? It is to Zoology and Botany that we turn to seek an answer to these questions; and in comparing the animals and plants of the Azores with those of other countries, we cannot shut out, even if we wished it, questions involving a theory of development as regards living productions.

On looking elsewhere on the surface of the globe, we find islands which, there is every reason to believe, were formerly joined to the nearest continent, and, on the other hand, we see islands which, there can be little doubt, have never existed but as islands. Under which of these two categories do the Azores come? Of the first class of islands we may mention Great Britain or Borneo.

The relationship of the fauna of the last-mentioned island to that of the adjoining continent of Asia has been recently ably investigated by Mr. Wallace, to whose instructive volumes on the Malay archipelago I refer my readers. Here we see that the salient features of "continental" islands are a flora fully representing all the elements of vegetable life found on the adjoining mainland, a fauna comprehending terrestrial Mammals as well as Reptiles and Amphibians, close connexion being also exhibited in other branches of zoology.

The characteristics of oceanic islands as regards their productions are chiefly negative, and may be briefly described as exhibiting poverty of species in general, remarkable absence of large groups of Plants, the absence of terrestrial Mammals (except Bats), Reptiles, and Amphibians.

Now, comparing the productions of the Azores with those of these two classes of islands, the inclination is decidedly towards the latter. There are no Mammalia, except one Bat, and neither Reptilia nor Amphibia; both the fauna and flora are decidedly poor in species, and many large groups of continental genera are unrepresented. Still there are, in certain classes of animals and in plants, some few species which are peculiar to the islands, and thus present the main difficulty in at once disposing of the question as to whether the Azores have always existed as oceanic islands or not.

The late Professor Edward Forbes endeavoured to account for the character of the fauna and flora of these islands by supposing them at one time to have been connected with the Madeiras and Canaries and with the south-western portion of Europe, arguing that the animal and vegetable productions, as we now find them, are the remains of what was once common to this old continent. The objections to this theory are numerous and, in my opinion, of great weight. In the first place, if it is necessary to join the Azores with what is now part of Europe to account for the European element of their fauna and flora, it is also necessary to join them with America to account for the American element; for it will not do to explain the presence of a few American forms by supposing them to have been introduced by accidental means and to require nothing less than a whole continent to explain the presence of the numerous European forms. The observed facts relating to insects and birds being found wandering at sea far from land, and of seeds being carried hundreds of miles on the surface of the ocean, and still germinating when brought to land, must be explained away; for, the fact of the accidental introduction of a species once admitted, the rest becomes merely a question of degree.

Is it possible that all traces of the elements of a "continental" fauna have been utterly eliminated from the Azores? Can this old continent have been thus destitute of Mammalia, or such an overwhelming catastrophe have overtaken the individuals that crowded on these peaks as to have destroyed them and every trace of their former existence?

The case of the Reptiles and the Freshwater Fish is nearly the same. There are no indigenous species of either class, with the single doubtful exception of *Lacerta dugesi* now found in Graciosa.

In the case of the Birds we see the process of immigration actually going on under our eyes. Seldom does a year pass but some stragglers are brought to the shores during the storms which so frequently prevail. I fully believe that the Wheatear has only very recently become established in this way in Corvo.

As regards the distribution of the species over the three groups of islands, we see that there is a numerical preponderance in the Eastern group, the Middle cluster containing a greater number than the Western. A fair induction from this fact is that the Eastern group, being nearer to the source of supply, has caught a greater number of stragglers than the Central, and the Central, for the same reason, more than the Western. Carrying out this observation, to include the Madeiras, we find a much greater numerical superiority in that archipelago over the Azores, and the same applies to the Canaries.

If climatal conditions are not to be taken into consideration when investigating the origin of the fauna and flora of a group of islands and comparing these productions with those of an adjacent continent, ought we not to see some analogy between the animal and vegetable life of groups of islands similarly situated, with respect to the nearest land, in different portions of the globe? In the case of the Azores we can fortunately find a parallel group of islands, the Galapagos, a glance at the Birds of which will give an excellent opportunity of testing how far a fauna is influenced in its component species by external causes. The avifauna of the Galapagos ought to bear to that of the adjacent continent of South America a relationship similar to that which the birds of the Azores do to those of the continent of Europe. What do we find? Beyond a general relationship and the identity of a few species, the birds of the Galapagos bear no specific resemblance to those of the continent of South America. These islands possess several peculiar genera; and in the cases where genera are common to the islands and the continent, the species are almost universally distinct. And further, almost every island possesses peculiarities in itself, a number of them being occupied by species which in the adjoining islands are represented by close allies, which can always be recognized as distinct. In fact the contrast could hardly be greater; and only when we look to the different conditions of climate do we begin to see that these must in a great measure have acted so as to produce the wide distinctions observable.

The Azores, almost entirely volcanic in their origin, are situated in a tempestuous sea, violent and constant storms blowing from all points of the compass; the climate is very humid and rainy; and currents flow in varying directions round and between the islands, apparently influenced by the wind prevalent at the time.

The Galapagos, also almost entirely volcanic in their origin, have a much drier climate, and enjoy almost uninterrupted calm; but between the islands flow strong unvarying currents. The results are apparent. The storms which rage around the Azores yearly bring straggling birds to their shores, not only supplying fresh colonists, but also serving to keep up the character of such species as have become already established. The calm atmosphere and deep unvarying currents in the Galapagian seas, on the other hand, not only sustain the distinctness of the island-forms from those of the mainland of America, but almost preclude the possibility of intercommunication between the different islands of the group.

To account for the indigenous productions of the Azores on the view that the islands have always been oceanic, it is necessary to begin at the beginning. Both in animals and vegetables we find peculiar features stamped upon a certain proportion of individual species. If this state of things has been brought about by the accumulation of slight modifications, time is required for the working of the process, and therefore we must turn to geology. This subject, as regards the Azores, has, so far as I am aware, been investigated by Hartung alone, according to whom the only stratified rocks (containing fossils of salt-water Mollusca) belong to the upper miocene period, and are only found in Santa Maria, an island I did not visit. All the other islands are of volcanic origin, the greater portion of the volcanos being now extinct; a few, however, still break out in periodic eruption. According to Lyell ('Elements of Geology,' 6th edit. p. 669), these volcanic eruptions commenced in the Upper Miocene period and continued down to Postpliocene times. Evidence is brought forward, from the presence of fragments of extraneous rock in Terceira* and Santa Maria, that the islands, or at least some of them, were nearly in their present state, as regards their area, during a subsequent glacial period. Thus the geological evidence brought to bear upon the date of the existence of land at this particular point of the earth's surface tends to show that the islands emerged at the close of the Miocene times, but long anterior to the Glacial period, and that previously to and since this latter epoch they have occupied much the same position, as regards contour and area, that they do at present. This determination of the date of the emergence of these islands, whether as islands or as a portion of a larger area, is of importance as showing that they could have received colonists of the Miocene flora which then prevailed in Southern Europe.

Geology, however, will throw but little light upon the vexed question as to the means by which these oceanic islands have received their faunas and floras. The only necessary requirement, if we adopt the theory of the gradual modification of different species, is that time enough should have existed to account for the greatest amount of change observable in any one species; and it would appear that in the Azores we have sufficient time for this purpose.

It is of course necessary to this view to suppose that the

* I searched the shore at Praia in Terceira with especial reference to the extraneous deposits mentioned by Hartung, and examined the walls into which some of the stones he mentions had been built. The impression I received from their inspection was that they were not in greater quantity nor larger than could be accounted for by supposing them to have been cast-away ballast. In former times Praia was a seaport of some importance; hence it is more probable that my suggestion may be the correct one than at first appears, the shipping-station having been moved to Angra. Similar stones are said to be more numerous at Villa do Porto in Santa Maria; but I made no personal inspection of them.

descendants of a common stock have survived in each locality to the present time. There is, however, another element which, if the theory of evolution is true, has a real existence, namely that the rate of change in any group is a very variable quantity: in some cases it might be great, and in a given time cause the species to present a great diversity when compared with individuals from the parent stock; in others, on the contrary, the rate of change might be very slow, and at the end of the same period they would differ slightly or not at all from those whence they were derived. Other circumstances, too, seem materially to bear upon the amount of difference observable between some Azorean species and their continental allies, supposing the islands to have been tenanted through the intervention of accidental causes. These causes have been always at work, and at different intervals have introduced species which have established themselves, some at remote times, others comparatively recently; and these species would therefore have periods of different lengths during which to accumulate modifications. In order that a species may become permanently established, it is only necessary that a pair should coexist and that they should find sufficient food for themselves and their young. Then, supposing a species to have become thus established, and that a fresh supply of individuals still kept arriving, the tendency would be to check the amount of variation and to keep up the likeness between the inhabitants of the islands and their continental brethren.

If, on the other hand, the requirements of the locality and circumstances induce a rapid rate of change in the first colonists, and hence the speedy *segregation* of an allied species, fresh immigrations of the original stock, unless arriving in a much increased numerical proportion, would be absorbed, or, unable to withstand competition, would die at once. Whenever a permanent difference had become established, the later immigrants might constitute a new colony of the parent stock, leaving the isolated form to continue its independent modifications. Thus we see how two or more distinct races, each derived from the same ancestry, but each owing its existence to a different combination of circumstances, might coexist in the same area.

It must also be borne in mind that the original stock in its own domain has not continued entirely stationary, but has also changed, at probably a slower rate, and in all probability in another direction, the surrounding influences being different.

In some cases the continental species, or parent stock, seems to have died out, leaving island colonists a remarkable feature in the fauna and flora when compared with the present productions of the continent whence they were originally derived.

Having thus endeavoured to sketch the kind of process by which differences of greater or less degree can be produced in the same or in different periods of time, I will now give a few instances where insects and birds have been found wandering in what I may call the Azorean seas. Such facts have an important and direct bearing upon the method by which oceanic islands may become tenanted by animal and vegetable life.

The captain of the whaler in which I took a passage from Fayal to Flores told me that a year or two before, as he was cruising about 400 miles to the south-west of the Azores, a white butterfly flew on board, which he caught and shut up in a small drawer in his cabin; when he opened the drawer again some time afterwards, he found that the insect had laid several eggs. He told me that he was coming from the south, from the Island of Ascension; so that it was not possible for the insect to have come on board in any European port and remained unobserved. He could not tell me the name of the butterfly, but *Pieris brassicæ* is abundant in the Azores. Had this one individual alighted on some island unoccupied by the species, and found a suitable plant whereon to deposit its eggs, the species might easily have become established from this one example.

In the 'Proceedings of the Zoological Society' for 1866 (p. 305) Mr. W. H. Flower records an instance of *Acherontia atropos* being taken at sea by the captain of the ship 'Hotspur,' during her homeward voyage, in lat. 40° 29′ N., long. 15° W., 260 miles from the coast of Portugal; also a specimen of *Sphinx convolvuli*, ♀, which flew on board the same ship in lat. 12° 9′ N., and long. 21° 17′ W.,—the prevailing winds being westerly and northerly: both these species are found in the Azores.

Mr. Salvin tells me that he saw several Swallows (*Hirundo rustica*) in one of his voyages to the West Indies, about 180 miles to the eastward of the Azores. This species is said to occur as an occasional straggler. I did not meet with it myself.

I also possess a skin of a Water-Rail (*Rallus aquaticus*) which was taken in lat. 46° 48′ N., long. 11° 30′ W., by the late Mr. W. Osburn when on his voyage to Jamaica in October 1867.

In a collection of bird-skins made in the China seas, and placed in the hands of Mr. Cuthbert Collingwood, is a specimen of a male Kestrel (*Tinnunculus alaudarius*), which was taken at sea in lat. 9° N., long. 28° W., in November 1863.

One cannot account for any of these instances by supposing the birds to have been in the act of migrating, as they were all observed out of their ordinary line. I have no doubt whatever that a very large number of cases of this sort, where insects and birds are found at sea and far from any land, might be placed on record if only the captains of vessels would interest themselves in the matter.

Before proceeding further with this question, let us analyze the groups of Azorean animals and plants concerning which our knowledge seems to be sufficiently advanced, and see what proportion of them are also to be met with on the continent of Europe, the Madeiras, and Canaries, and also what proportion of each group may be considered peculiar to the islands.

As stated above, Mammalia, Reptilia, and Amphibia are unrepresented, so also are Freshwater Fish. Our knowledge of Nocturnal Lepidoptera is much too incomplete to give any trustworthy results; the same may be said of the other orders of insects, except Coleoptera and Diurnal Lepidoptera. The Mosses and Hepaticæ, too, have hardly been sufficiently investigated to help us much; and this portion of the subject having been somewhat differently treated by Mr. Mitten, I think it advisable for the present to leave them out of my calculation.

We have, then, to consider the distribution of the Birds. Diurnal Lepidoptera, Coleoptera, Land and Freshwater Mollusca, and the Plants, except the Mosses and Hepaticæ. These are distributed numerically as follows:—

	Total number of species.	Percentage of species: Common to the Azores and Europe.	Madeiras.	Canaries.	Peculiar to the Azores.	
Aves	53	91	75	57	2	1 American.
Diurnal Lepidoptera	8	87	...	...	0	1 American.
Coleoptera	212	83	65	55	6	3 American.
Land and Freshwater Mollusca	69	38	10	6	46	
Plants	480	83	62½	54	8¼	5 American and African.

We thus see that a very remarkable uniformity exists as

regards the distribution numerically of the species of Birds, Coleoptera, and Plants. The comparison shows that between 80 and 90 per cent. of the species of these groups are also found in Europe, that about 65 per cent. are also found in Madeira, and about 55 per cent. are also Canarian, and that numbers varying from 2 to 8¼ per cent. are not found out of the islands. As regards the Mollusca, it appears that the amount of peculiarity is much greater, reaching to 46 per cent. of the whole number; but I think this result must be received with caution, as it is the habit of conchologists to take a much closer view of species (and that, too, from an examination of the shell only) than prevails among naturalists who devote themselves to other branches. Moreover Mr. Tristram remarks that he conceives that 27 of the 32 peculiar species would be admitted to be more closely allied to European types than to any other. It would be, perhaps, incorrect to consider these 27 species insufficiently distinct from their European allies and to treat them as mere varieties; but thus considered we find that the Mollusca conform to the numerical distribution of the above-named groups so far as the European and Azorean species are concerned. Passing the 27 species to the European heading of the account, we find that of the 69 species 77 per cent. are also European, and 7 per cent. peculiar.

The American element in the fauna and flora is very slight. In the case of the birds, one only (*Thalassidroma wilsoni*), and that an oceanic wanderer in the North Atlantic, can, hardly, be said to connect the Azores and America. Among the Diurnal Lepidoptera, the *Danais* can scarcely be considered an established resident, though its accidental presence in two of the islands is remarkable. Three species of Coleoptera are also Brazilian, and were probably introduced; and of the Plants four species (*Lepidium virginicum*,

Cakile americana, Cyperus vegetus, Lycopodium cernuum) are common to the Azores and the American continent.

The connexion between the Azores and the Ethiopian region is slighter still. One peculiar species of Beetle (*Elastrus dolosus*, Jard.) has its nearest known ally in Madagascar; and one plant (*Myrsine africana*) is identical with a species ranging over Intertropical and South Africa.

The Azorean fauna and flora thus resolves itself into the following elements:—

(1) Species identical with European.

(2) Species represented generically in Europe.

(3) Species unrepresented in Europe either specifically or generically.

(4) Species not European, but found in America or elsewhere.

It remains, then, to show how one may account for the origin of these different elements on the supposition that the islands have been continuously stocked by European colonists from the close of the Miocene period down to the present time.

(1) The species which are identical with those now existing in Europe may be considered the most recent immigrants, or if they have arrived at a more distant period, divergence from the parent stock has either not progressed, or has been kept in check by a stream of colonists sufficiently large to maintain the inhabitants of the islands and the continent specifically true.

(2) In the case where a genus is represented in Europe and in the Azores by different species, we must place the date of the arrival of the first colonists at a more remote period, since which time checks upon the tendency to diverge have not been sufficiently strong to hinder the establishment of a different race.

(3) The highly peculiar forms which, though found in

the Azores, are unrepresented in Europe at the present time, must have the date of their introduction thrown back to that far distant epoch when the fauna and flora of the continent bore the impress which prevailed in Miocene times, but which, in Europe, at least, has since been almost entirely obliterated.

(4) The element which is not traceable to Europe either in recent or past times is exceedingly small, and must be attributed to immigration from other parts of the world at no distant date.

It is not to be supposed that we shall ever be able to trace accurately the outlines of the past history of any species, to say nothing of the higher groups into which our science constrains us to classify the organic world. But I believe that by multiplying the record of facts relating to the distribution of species, an approximate knowledge of the origin of the salient features of many a fauna and flora is attainable, and that more light will be continually thrown upon the connexion, not only between the organisms of islands and continents, but also between those of present and past times.

It is to the manner in which animals and plants are distributed over the face of the globe that we must also look for an ever increasing weight of argument having a direct bearing upon the theory of the derivative origin of species.

INDEX TO SCIENTIFIC NAMES.

THE END.

Printed by TAYLOR and FRANCIS, Red Lion Court, Fleet Street.

Made in United States
North Haven, CT
11 November 2026

22197231R00202